Users Guide to Ecohydraulic Modelling and Experimentation

IAHR Design Manual

Series editor

Peter A Davies
Department of Civil Engineering,
The University of Dundee,
Dundee,
United Kingdom

The International Association for Hydro-Environment Engineering and Research (IAHR), founded in 1935, is a worldwide independent organisation of engineers and water specialists working in fields related to hydraulics and its practical application. Activities range from river and maritime hydraulics to water resources development and eco-hydraulics, through to ice engineering, hydroinformatics and continuing education and training. IAHR stimulates and promotes both research and its application, and, by doing so, strives to contribute to sustainable development, the optimisation of world water resources management and industrial flow processes. IAHR accomplishes its goals by a wide variety of member activities including; the establishment of working groups, congresses, specialty conferences, workshops, short courses; the commissioning and publication of journals, monographs and edited conference proceedings; involvement in international programmes such as UNESCO, WMO, IDNDR, GWP, ICSU, The World Water Forum; and by co-operation with other water-related (inter)national organisations.
www.iahr.org

Users Guide to Ecohydraulic Modelling and Experimentation

Experience of the Ecohydraulic Research Team (PISCES) of the HYDRALAB Network

Editors

L.E. Frostick & R.E. Thomas
Department of Geography, Environment and Earth Sciences, University of Hull, Hull, UK

M.F. Johnson & S.P. Rice
Department of Geography, Loughborough University, Loughborough, UK

S.J. McLelland
Department of Geography, Environment and Earth Sciences, University of Hull, Hull, UK

CRC Press
Taylor & Francis Group
Boca Raton London New York Leiden

CRC Press is an imprint of the
Taylor & Francis Group, an **informa** business

A BALKEMA BOOK

Acknowledgement:
The work described in this publication was supported by the European Community's Seventh Framework Programme through the grant to the budget of the Integrated Infrastructure Initiative HYDRALAB IV, Contract no 261520 (RII3).

Please refer to this publication as follows:
Frostick, L.E.; Thomas, R.E.; Johnson, M.F.; Rice, S.P. & McLelland, S.J. (eds) 2014. Users Guide to Ecohydraulic Modelling and Experimentation: Experience of the Ecohydraulic Research Team (PISCES) of the HYDRALAB Network, CRC Press/Balkema, Leiden, The Netherlands.

Typeset by V Publishing Solutions Pvt Ltd., Chennai, India
Printed and Bound by CPI Group (UK) Ltd, Croydon, CR0 4YY

Although all care is taken to ensure integrity and the quality of this publication and the information herein, no responsibility is assumed by the publishers nor the author for any damage to the property or persons as a result of operation or use of this publication and/or the information contained herein.

Published by: CRC Press/Balkema
                P.O. Box 11320, 2301 EH Leiden, The Netherlands
                e-mail: Pub.NL@taylorandfrancis.com
                www.crcpress.com – www.taylorandfrancis.com

CRC Press/Balkema is an imprint of the Taylor & Francis Group, an informa business

Library of Congress Cataloging-in-Publication Data

applied for

ISBN: 978-1-138-00160-2 (Hbk)
ISBN: 978-1-315-77883-9 (eBook PDF)

About the IAHR Book Series

An important function of any large international organisation representing the research, educational and practical components of its wide and varied membership is to disseminate the best elements of its discipline through learned works, specialised research publications and timely reviews. IAHR is particularly well-served in this regard by its flagship journals and by the extensive and wide body of substantive historical and reflective books that have been published through its auspices over the years. The IAHR Book Series is an initiative of IAHR, in partnership with CRCPress/Balkema – Taylor & Francis Group, aimed at presenting the state-of-the-art in themes relating to all areas of hydro-environment engineering and research.

The Book Series will assist researchers and professionals working in research and practice by bridging the knowledge gap and by improving knowledge transfer among groups involved in research, education and development. This Book Series includes Design Manuals and Monographs. The Design Manuals contain practical works, theory applied to practice based on multi-authors' work; the Monographs cover reference works, theoretical and state of the art works.

The first and one of the most successful IAHR publications was the influential book *"Turbulence Models and their Application in Hydraulics"* by W. Rodi, first published in 1984 by Balkema. I. Nezu's book *"Turbulence in Open Channel Flows"*, also published by Balkema (in 1993), had an important impact on the field and, during the period 2000–2010, further authoritative texts (published directly by IAHR) included *Fluvial Hydraulics* by S. Yalin and A. Da Silva and *Hydraulicians in Europe* by W Hager. All of these publications continue to strengthen the reach of IAHR and to serve as important intellectual reference points for the Association.

Since 2011, the Book Series is once again a partnership between CRCPress/Balkema – Taylor & Francis Group and the Technical Committees of IAHR and I look forward to helping bring to the global hydro-environment engineering and research an exciting set of reference books that showcase the expertise within the IAHR Community.

Peter A Davies
University of Dundee, UK
(Series Editor)

Contents

Foreword

Many recent studies have shown that hydrodynamics controls most biophysical processes in aquatic ecosystems leading to the emergence of a quickly growing branch of fluid mechanics known as ecohydraulics or ecohydrodynamics. This new discipline is still emerging at the interfaces between fluid mechanics, ecology, biology, and biomechanics, demonstrating a multidisciplinary approach in modern science. Although the importance of ecohydraulics for developing better management strategies for aquatic ecosystems is widely recognised, the actual knowledge of hydrodynamic effects in flow-biota interactions remains very limited.

Slow progress in this field is particularly due to measurement difficulties at scales most relevant to organisms, poorly understood biomechanical properties of organisms that change significantly across species, scales, and environments, and the absence of a sound unifying interdisciplinary platform for integrating and upscaling hydrodynamic, biomechanical and ecological processes. In recent years, however, these negative factors have been significantly weakened with the appearance of new measurement capabilities (e.g., high-resolution acoustic and optical instrumentation), employment of modern hydrodynamic concepts in ecohydraulic studies, and advances in biomechanics. The information related to these important developments is dispersed in a variety of ecological, biological, geomorphological, engineering, and physical journals and conference proceedings. The uptake of this information by a wider research community encounters, therefore, multiple challenges due to different research cultures, discipline languages, and expectations, which vary significantly between fluid mechanics, ecology, and biomechanics. The integration of this information within a single book and at a level suitable for engineers, physicists, geomorphologists, biologists, and ecologists is a mammoth task, which one may consider simply impossible. This book, however, proves the opposite. Without doubt, it represents a significant step forward in advancing ecohydraulics and, even more importantly, in consolidating research efforts of such diverse research communities. This outstanding result reflects strong interdisciplinary authorship and significant dedication to the goal of this book. The authors are all internationally leading experts who themselves largely pushed the boundaries of this field forward, making reading even more thrilling. The book thus responds to the growing demands for advanced ecohydraulic knowledge in numerous applications, including civil and environmental engineering, resource management, aquaculture, bioenergy technologies, and bio-security.

Although the book title sounds quite specialised, i.e., "Users Guide to Ecohydraulic Modelling and Experimentation", its scope is wider and an interested reader will find there much more than just recipes for experimentation. The book structure is well thought out and includes two parts. The first part covers general issues of measurement methodologies, facilities, instrumentation, and organism handling; while the second part is focused on biofilms, plants, and macrozoobenthos, demonstrating the power of experimentation in developing new knowledge of complex flow-organisms interactions. In this respect the book is unique as it may serve as both a brief introduction to ecohydraulics and as a guide to experimentalists, for the first time combining in-depth considerations of how to plan, design, conduct, and interpret ecohydraulic experiments. Currently, there are significant differences in experimental approaches, facilities, instrumentation, and data analysis routines across disciplines that often make data comparisons and generalisations in ecohydraulics very difficult. This book addresses this problem by providing a common platform for standardisation of experimental studies in ecohydraulics that is currently missing but urgently needed.

Early career researchers and PhD students will particularly benefit from this book that will help them to make the first steps in their research in a more comprehensive (and speedy!!) way. Experienced researchers will find in this book the answers to questions that have been already addressed by their colleagues working in a different discipline. In fact, ecologists are often unaware about the work of hydraulic researchers and vice versa, and this book will help bridge and consolidate different communities involved in studies of aquatic ecosystems.

The book offers great opportunities to researchers and professionals with different backgrounds to "synchronise" their efforts in addressing multiple knowledge gaps at the borders between fluid mechanics, biomechanics, and ecology, i.e., areas where the probability of new discoveries is highest. The obtained experimental information will provide a missing platform for developing process-based models that should replace currently used empirical approaches, which are often operating with coefficients disconnected to the underlying processes and actual organisms.

The Editors and the Authors should be highly commended for their efforts that resulted in such an excellent book. It gives me great pleasure and privilege to recommend it to everybody dealing with ecohydraulic problems. I am sure many readers will enjoy reading this book and will join me in congratulating the authors on this great accomplishment.

Professor Vladimir Nikora, FRSE
Sixth Century Chair in
Environmental Fluid Mechanics
University of Aberdeen, UK

List of symbols and abbreviations

Symbol	Unit	Meaning
ADV		Acoustic Doppler Velocimetry
ADVP		Acoustic Doppler Velocity Profiling
AFDM		Ash Free Dry Mass
CLSM		Confocal Laser Scanning Microscopy
DANS		Double-Averaged Navier-Stokes equations
DEM		Digital Elevation Model
DFS		Drag Force Measurement System
DM		Dried Matter
DNS		Direct Numerical Simulation
DoD		DEM of Difference
EMS		Electromagnetic Sensor
EPS		Extracellular Polymeric Substance
HWA		Hot-Wire/Hot-Film Anemometry
Hz		Hertz
KC		Keulegan–Carpenter number
LAI		Leaf Area Index
LDA		Laser Doppler Anemometry
LDV		Laser Doppler Velocimetry
LES		Large Eddy Simulation
PIV		Particle Image Velocimetry/Velocimeter
RANS		Reynolds-Averaged Navier-Stokes
RD		Roughness Density
S		Signal Amplitude
SDG		Submersible Drag Gauge
SNR		Signal-to-Noise Ratio
SRB		Sulphate Reducing Bacteria
SRP		Soluble Reactive Phosphorous
SSC		Suspended Sediment Concentration
TKE		Turbulent Kinetic Energy
UDVP		Ultrasonic Doppler Velocity Profiling

A_0	m²	Area of the water-bed interface bounded by an averaging domain
AC_D		Area-drag coefficient product
A_p	m²	Frontal (projected) area of a plant per unit volume
c	m s⁻¹	Speed of sound
C_D		Drag coefficient
D_{50}	m	Median grain-size
D_b		Bioturbation coefficient
d_s	m	Stem diameter
E	N m⁻²	Young's modulus of elasticity
f_0		Frequency of the transmitted sound
f		Sampling frequency
F	N m⁻²	Force
f_D		Change in received frequency (Doppler-shift frequency)
F_D	N m⁻²	Drag force
ff		Darcy-Weisbach roughness coefficient
Fr		Froude number
g	m s⁻²	Gravitational acceleration
I [cap I]		Moment of intertia
J	N m²	Flexural stiffness
K		Permeability of a porous medium
k		The dissipation length scale
k_s		Equivalent hydraulic roughness
l [small L]		Boundary layer thickness
L	m	Length
L_e		Eddy length scale
L_s	m	Stem length
M		Bending stiffness
N	dB	Background noise level
P	Pa	Pressure
P_{Max}		Standardised maximum photosynthetic rate
pH		Level of acidity/alkalinity
Q	m³ s⁻¹	Flow discharge
R^2		Radial velocities
Re		Reynolds number
Re_v		Stem Reynolds number
S_f		Friction slope
t	s	Time
T	s	Sampling time period
T_p	s	Peak wave period
u, v, and w	m s⁻¹	Orthogonal velocity components
V_0	m³	Volume
X		Linear length scale
y		Distance between suspended particles and transducer
Y		Non-linear momentum loss coefficient
z_0	m	Roughness length

α		Autocorrelation coefficient
δ	m	Deflection of the stem
ε		Dissipation rate per unit mass
κ		von Kàrmàn constant
λ		Smallest timescale of flow
μ	Pa s	Dynamic viscosity
∂		Partial differential operator
ρ	Kg m^{-3}	Density
ρ_t	Kg m^{-3}	Tissue density
ρ_w	Kg m^{-3}	Mass density of water
ν	m^2 s^{-1}	Kinematic viscosity
ϕ		Porosity
φ		Dimensionless spectral width

Periodic table of the elements

1A	2A	3B	4B	5B	6B	7B	8B	8B	8B	1B	2B	3A	4A	5A	6A	7A	8A
1 H 1.00794 Hydrogen																	2 He 4.002602 Helium
3 Li 6.941 Lithium	4 Be 9.012182 Beryllium											5 B 10.811 Boron	6 C 12.0107 Carbon	7 N 14.0067 Nitrogen	8 O 15.9994 Oxygen	9 F 18.9984032 Fluorine	10 Ne 20.1797 Neon
11 Na 22.989769 Sodium	12 Mg 24.3050 Magnesium	3B 21 Sc 44.955912 Scandium										13 Al 26.9815386 Aluminum	14 Si 28.0855 Silicon	15 P 30.973762 Phosphorus	16 S 32.065 Sulfur	17 Cl 35.453 Chlorine	18 Ar 39.948 Argon
19 K 39.0983 Potassium	20 Ca 40.078 Calcium	21 Sc 44.955912 Scandium	22 Ti 47.867 Titanium	23 V 50.9415 Vanadium	24 Cr 51.9961 Chromium	25 Mn 54.938045 Manganese	26 Fe 55.845 Iron	27 Co 58.933195 Cobalt	28 Ni 58.6934 Nickel	29 Cu 63.546 Copper	30 Zn 65.38 Zinc	31 Ga 69.723 Gallium	32 Ge 72.64 Germanium	33 As 74.92160 Arsenic	34 Se 78.96 Selenium	35 Br 79.904 Bromine	36 Kr 83.798 Krypton
37 Rb 85.4678 Rubidium	38 Sr 87.62 Strontium	39 Y 88.90585 Yttrium	40 Zr 91.224 Zirconium	41 Nb 92.90638 Niobium	42 Mo 95.96 Molybdenum	43 Tc [98] Technetium	44 Ru 101.07 Ruthenium	45 Rh 102.90550 Rhodium	46 Pd 106.42 Palladium	47 Ag 107.8682 Silver	48 Cd 112.411 Cadmium	49 In 114.818 Indium	50 Sn 118.710 Tin	51 Sb 121.760 Antimony	52 Te 127.60 Tellurium	53 I 126.90447 Iodine	54 Xe 131.293 Xenon
55 Cs 132.9054519 Cesium	56 Ba 137.327 Barium	57-71 Lanthanides	72 Hf 178.49 Hafnium	73 Ta 180.94788 Tantalum	74 W 183.84 Tungsten	75 Re 186.207 Rhenium	76 Os 190.23 Osmium	77 Ir 192.217 Iridium	78 Pt 195.084 Platinum	79 Au 196.966569 Gold	80 Hg 200.59 Mercury	81 Tl 204.3833 Thallium	82 Pb 207.2 Lead	83 Bi 208.98040 Bismuth	84 Po [209] Polonium	85 At [210] Astatine	86 Rn [222] Radon
87 Fr [223] Francium	88 Ra [226] Radium	89-103 Actinides	104 Rf [267] Rutherfordium	105 Db [268] Dubnium	106 Sg [271] Seaborgium	107 Bh [272] Bohrium	108 Hs [270] Hassium	109 Mt [276] Meitnerium	110 Ds [281] Darmstadtium	111 Rg [280] Roentgenium	112 Cn [285] Copernicium	113 Uut [284] Ununtrium	114 Fl [289] Flerovium	115 Uup [288] Ununpentium	116 Lv [293] Livermorium	117 Uus [294] Ununseptium	118 Uuo [294] Ununoctium

Lanthanides														
57 La 138.90547 Lanthanum	58 Ce 140.116 Cerium	59 Pr 140.90765 Praseodymium	60 Nd 144.242 Neodymium	61 Pm [145] Promethium	62 Sm 150.36 Samarium	63 Eu 151.964 Europium	64 Gd 157.25 Gadolinium	65 Tb 158.92535 Terbium	66 Dy 162.500 Dysprosium	67 Ho 164.93032 Holmium	68 Er 167.259 Erbium	69 Tm 168.93421 Thulium	70 Yb 173.054 Ytterbium	71 Lu 174.9668 Lutetium

Actinides														
89 Ac [227] Actinium	90 Th 232.03806 Thorium	91 Pa 231.03588 Protactinium	92 U 238.02891 Uranium	93 Np [237] Neptunium	94 Pu [244] Plutonium	95 Am [243] Americium	96 Cm [247] Curium	97 Bk [247] Berkelium	98 Cf [251] Californium	99 Es [252] Einsteinium	100 Fm [257] Fermium	101 Md [258] Mendelevium	102 No [259] Nobelium	103 Lr [262] Lawrencium

Contributors

HYDRALAB Partner/Contributor	email
Centre National de la Recherche Scientifique (CNRS)	
Olivier Eiff	Olivier.Eiff@imft.fr
Fredéric Moulin	Frederic.Moulin@imft.fr
Deltares	
Jasper Dijkstra	Jasper.Dijkstra@deltares.nl
Ellis Penning	Ellis.Penning@deltares.nl
Gottfried Wilhelm Leibniz University Hannover (LUH)	
Maike Paul	paul@fzk-nth.de
Michalis Vousdoukas	vousdoukas@fzk-nth.de
Loughborough University (ULBORO)	
Matt Johnson	M.F.Johnson@lboro.ac.uk
Stephen Rice	S.Rice@lboro.ac.uk
University of Hull (UHULL)	
Lynne Frostick	L.E.Frostick@hull.ac.uk
Stuart McLelland	S.J.Mclelland@hull.ac.uk
Dan Parsons	D.Parsons@hull.ac.uk
Martyn Pedley	H.M.Pedley@hull.ac.uk
Mike Rogerson	M.Rogerson@hull.ac.uk
Paul Saunders	Paul.Saunders@hull.ac.uk
Robert Thomas	R.Thomas@hull.ac.uk

Acknowledgements

The editors would like to thank Linda Love for all the hard work she has put into the final editing of this book. Without her the manuscript would not have been completed to such a high standard and in a timely manner. We would also like to thank Simone van Schijndel for her encouragement and Peter Davies for his support in converting the work into an IAHR publication. We are grateful to Vladimir Nikora for his very positive response to an earlier draft of the book and for agreeing to write a foreword for us.

The work described in this publication was supported by the European Community's 7th Framework Programme through the grant to the budget of the Integrated Infrastructure Initiative HYDRALAB-IV, Contract no. 261520.

Chapter 1

Introduction

L.E. Frostick, M.F. Johnson, R.E. Thomas,
S.P. Rice & S.J. McLelland

1.1 INTRODUCTION

The complex interactions between aquatic organisms and physico-chemical processes are fundamental to our understanding and management of both individual biotic communities and the total environment. Of significance are both how organisms modify their environment and how they and their behaviour are, in turn, modified by the environment around them. In marine and freshwater aquatic systems, the interactions between organisms and the prevailing hydraulic conditions are of prime importance. However, the field of experimental and numerical hydraulic modelling frequently reduces the complex consequences of plant and animal interactions with flows to a passive effect on bed roughness. Few experiments in hydraulic facilities have used prototype organisms. Some exceptions are: Statzner *et al.* (2006, crayfish), Johnson *et al.* (2011, crayfish), Wilson *et al.* (2003, vegetation), Fathi-Moghadam & Kouwen (1997, vegetation), Battin (2003, biofilms). This is often because experimental hydraulic facilities are predominantly designed without consideration of the need to keep organisms both alive and behaving as they would in the natural environment. Some ecological experiments have attempted to use experimental hydraulic facilities to model the active responses of organisms to flows. However, many of these experiments are designed with little knowledge of the constraints on model set-up and flow measurement. The results of poorly designed experiments are unlikely to reflect processes in the environments being simulated and in some cases the data may be meaningless.

Organisms interact with physical processes and environments at all spatial and temporal scales. Burrowing by benthic invertebrates was responsible for the oxygenation of ocean sediments 500 million years ago and plants exert a control over the plan-form and morphology of rivers (Thayer, 1979; Gran & Paola, 2001). Removal or alteration of communities of organisms can have severe consequences for the functioning of environments by changing geomorphic, hydraulic and biogeochemical processes (Coleman & Williams, 2002). Research at the interface between biological and environmental studies has a long history. For example, Charles Darwin is credited with the first study of bioturbation, i.e. the burrowing of animals, and writing a book dedicated to the importance of earthworms in the formation of soil (Darwin, 1881). The theory of plant succession, described by Clement (1916), is largely based on the concept that plant communities alter the physical environment, benefiting later plant and animal colonists. The work of Fager (1964) and Rhoads & Young (1970a, 1970b) stimulated the study of bioturbation and biostabilisation in marine environments, which are now considered fundamental processes in the formation, morphology and functioning of marine, estuarine and lacustrine sediments. More recently, over the last twenty years, the development of new and improved techniques such as X-ray imaging, remote sensing, Acoustic Doppler Velocimetry (ADV) and Particle Image Velocimetry (PIV), has led to a reinvigoration of organism-environment studies and rapid advancement of our knowledge. This has also led to the appearance of numerous inter-disciplinary fields including "ecohydraulics", "biogeomorphology" and "ecological engineering".

Research at the interfaces among ecology, environment and hydraulic engineering requires knowledge of each of the disciplines. However, it can be challenging to access information across disciplinary boundaries because of the differing research methods, aims, and terminology. This can result in information from one discipline being inaccessible to others. Physical modelling using laboratory flume experiments provides a specific example where cross-disciplinary knowledge is essential (Rice et al., 2010a). Although the main benefit of this type of experimentation is the ability to simplify nature in order to isolate specific processes, over-simplification of the flow, sedimentary environment or organism ecology can lead to experiments that have no relevance to natural systems, leading to results that have limited validity or applicability. It is important to ensure that the laboratory environment is analogous to nature and that organisms behave in a manner that is analogous to their wild equivalents. Consequently, knowledge of the biology and physics of the environment is essential in the design of successful ecohydraulic flume experiments. However, many important considerations may be taken for granted by scientists in one discipline that may not be thought of by colleagues from another; for example, the necessity to feed animals or the length of measurement time necessary to quantify turbulence. Consequently, such considerations are not normally discussed in detail in academic publications. This makes it difficult for cross-disciplinary researchers to obtain the relevant information when designing experiments, particularly those involving living organisms and complex hydrodynamic environments. It is therefore essential for there to be interchange of experience as well as best-practice guidelines, such as those developed in HYDRALAB IV (Frostick et al., 2011).

1.2   RATIONALE FOR THE BOOK

This guide has been compiled by the interdisciplinary team of ecologists, geomorphologists, sedimentologists, hydraulicists, and engineers involved in HYDRALAB IV, the European Integrated Infrastructure Initiative on hydraulic experimentation which forms part of the European Community's Seventh Framework Programme. It is designed to give an overview of our current knowledge of organism-environment interactions in marine and freshwater aquatic systems and to provide guidance to those wishing to use hydraulic experimental facilities to explore ecohydraulic processes. By highlighting the current state of our knowledge, this design manual will act as a guide to the use of living organisms in physical models and help scientists and engineers understand limitations on the use of surrogates. It incorporates chapters on the general decisions that need to be taken when designing an ecohydraulic experiment as well as specific chapters on the main aquatic and marine organisms likely to be of interest. Each of the chapters reviews current knowledge in a defined area of ecohydraulic experimentation research.

1.3   REASONS FOR USING PHYSICAL MODELLING IN ECOHYDRAULICS

Although incorporating animals and plants into hydraulic experiments is challenging, it is an essential step in gaining insight into their importance for modifying flow and sediment transport in the natural environment and their response to given flow and/or sediment transport regimes. Physical models allow insight into phenomena that are as yet unknown or are sufficiently complicated that it is difficult or impossible to derive a numerical solution. Without such experiments to provide data for numerical models, it will be impossible to extrapolate the consequences of climate change, which is likely to result in sea level rise and increasing storminess, on flooding and coastal defence. Equally, biologists wishing to understand the forces acting on biota and the likelihood of survival under predicted climate change scenarios must experiment with the organisms of concern.

An additional advantage of using a physical modelling approach to ecohydraulic research is the capacity to directly observe organism behaviour and modify the experiments accordingly. This is only possible as a result of the high level of flow and measurement control possible in a laboratory flume or tank. However, care must be taken in the interpretation of the results as all physical modelling approaches incorporate elements of simplification and scaling and it is not always possible to model all relevant variables as they would exist in the natural environment (see Frostick *et al.*, 2011, section 1.3 for further discussion).

1.4   AUDIENCE FOR THE BOOK

This book is aimed at:

- Hydraulic scientists and engineers interested in the influence of organisms on currents and waves
- Sedimentologists concerned with the ways in which sediment erosion and deposition may be modified by organism growth and behaviour
- Environmental scientists interested in interactions amongst and between organisms and the physical environment
- Ecologists and biologists dealing with the behaviour distribution, dispersal, physiology and success of organisms in aquatic and marine environments.

It is intended to be used by both academic researchers and those in commercial laboratories. As it incorporates state-of-the-art technical details, research and ideas, it is aimed at experienced post-graduate level engineers and scientists and is not a student textbook.

1.5   CHOICE OF FACILITY

The choice of facility for a series of experiments is the important first step. A mistake at this stage can mean that the experiments are inappropriate for the research aims and the results will therefore be, at best, difficult to interpret and, at worst, useless. A general description of the types of facilities available is given in section 1.5 of Frostick *et al.* (2011) and will not be repeated here. However there are specific limitations which apply to ecohydraulic experimentation alone. The facility selection process must take these into account and must carefully consider both the scale and character of the ecohydraulic phenomenon being researched and the type of organisms involved. A range of different facilities are available for use but few large flumes and wave tanks have been designed with organisms in mind. For example, there are some facilities able to use salt water (e.g. the Total Environment Simulator at Hull, UK), but many cannot and this is a limitation on the planning of any experiments with marine organisms. The availability of lighting that provides the appropriate balance between light of different wavelengths and frequencies (e.g. grow lamps) is an important consideration for experiments incorporating photosynthetic plants. Similarly, the presence of chemical contaminants in the water may inhibit plant and animal growth and can lead to the poor health and death of the organisms. All of these considerations are discussed further in Chapter 2 of this book.

1.6   CHOICE OF MATERIALS

The choice of materials requires careful consideration and should be guided by the aims of the experiments. The first and most important decision is whether to use prototype plants and animals or surrogates. The problems associated with animal and plant husbandry can be circumvented if inanimate surrogates can be substituted for the prototype organisms without detriment to the relevance of the data collected. However, this may preclude complex interactions between biota and the physical environment which are not predictable yet still important. Once the decision is made to use either living organisms or surrogates, the questions of water quality and substrate character follow. Surrogates in general put fewer constraints on these materials, but again the aim of the experiments and the required quality of the data will control these decisions.

1.7   CHOICE OF MEASUREMENT EQUIPMENT

Measuring flow around living organisms introduces challenges that add further to the already difficult problems faced by hydraulic engineers in experimental situations. Most techniques are not suitable for measuring close to mobile boundaries and many introduce stresses that have the potential to alter the behaviour of the target organisms. Irrespective of the technique adopted, these constraints on selecting an appropriate suite of flow measurement techniques must be carefully considered and placed alongside the generic issues that impact upon the successful measurement of turbulent flows. Important amongst these is the crucial issue of selecting appropriate measurement duration. The technique selected will almost inevitably be a compromise between the requirements for excellent data and the constraints imposed by the organisms being studied.

1.8 LAYOUT OF THE BOOK

Each of the seven main chapters in this book deals with an important aspect of ecohydraulic experimentation in hydraulic infrastructures. The chapters are organised into two sections followed by a conclusion that summarises the important design criteria for experiments in ecohydraulics. Researchers are recommended to read all of Section 1 and then the part of Section 2 relevant to their model organisms. These are summarised below.

1.8.1 Methods, materials and measurement

This section outlines the decision-making process needed to plan and commission effective ecohydraulic experiments. It includes the following chapters:

Chapter 2 Maintaining the health and behavioural integrity of plants and animals in experimental facilities

This chapter focuses on the importance of animal and plant husbandry when using biota in hydraulic facilities. One of the prime considerations when carrying out experiments using living plants and animals is the ethical obligation to maintain organism health, reduce stress and ensure behaviour remains as close to that in the natural environment as possible. This may be a particular challenge to physical scientists and engineers more familiar with working with inert materials. Considerations of water chemistry and temperature and daylight availability are discussed. There are limits to the manipulation of environmental conditions in flume facilities and these are outlined so that biologists and ecologists interested in ecohydraulics can decide whether working in hydraulic infrastructures is appropriate for their research.

Chapter 3 Using surrogates, including scaling surrogates, in laboratory flumes and basins

Surrogates are approximate replicas of biota, mimicking specific organism traits of relevance to the aims and objectives of the study. Physical surrogates vary considerably in how closely they reproduce organism characteristics and this will constrain the utility of any results obtained. The aims and objectives of the experiments are therefore paramount when selecting and constructing surrogates. This chapter outlines the main types of surrogates that have been used as well as the process of scaling organism effects to fit the available facilities. There is a detailed analysis of the advantages and disadvantages of using surrogates rather than prototype organisms.

Chapter 4 Flow measurements around organisms and surrogates

This chapter reviews the difficult topic of selecting the method used to measure the relevant properties of a turbulent fluid flowing through and around flora and fauna. There is a brief discussion of turbulent flows and the general difficulties associated with obtaining meaningful measurements. The challenges inherent in measuring around mobile materials, whether waving vegetation or motile animals, are emphasised. Issues associated with the impact of using specific techniques on animal and plant health, particularly laser-based technology, are discussed.

1.8.2   Organism-specific considerations

Chapter 5 Biofilms, hydraulics and sediment dynamics

Biofilms are complex assemblages of bacteria, fungi and algae that develop on all submerged surfaces in aquatic systems. Even though they are ubiquitous, few experiments have explored their interplay with hydraulics and sediment transport. This chapter explores the character of biofilms, their diversity and the constraints on their growth and morphology. It focuses on the factors that might influence the development and viability of biofilms within the flume and wave tank environment. It also summarises current knowledge of biofilm-hydraulic interactions.

Chapter 6 Plants, hydraulics and sediment dynamics

Vegetation exerts an important influence on the flow and the morphology of aquatic systems. It influences flow resistance and turbulence and so changes transport patterns of sediment, nutrients, and contaminants. Vegetation is therefore an important focus for ecohydraulic research. This chapter explores how plants have been used in hydraulic experimentation and gives guidance on the advantages and disadvantages of the many different approaches to this type of research. It also touches on the importance of plant ecohydraulic knowledge in effective coastal and river management.

Chapter 7 Macrozoobenthos, hydraulics and sediment dynamics

This chapter reviews the impact of both mobile and sessile macrozoobenthos on physical processes in aquatic environments, focussing on the issues important to experimental research in hydraulic facilities. Macrozoobenthos include all animals that are greater than 1 mm in length. However, this book does not cover research into benthic or other fish, which are a separate and specialist area. This chapter focuses on research into the impacts of mobile animals on the topography and structure of surface and near-surface sediments. This is a popular area of research but much less is known of how this impacts on the hydrodynamic environment despite evidence to suggest that this may be large. It also reviews the interactions between sessile animals and the hydraulic environment. Sessile animals are those that attach themselves to substrates and, therefore, can remain immobile for extended periods. They modify the hydraulic environment as currents adjust to the presence of their bodies, increasing flow heterogeneity by creating areas of accelerating and decelerating flow and inducing drag and lift forces. Finally, there is a review of methods used in physical experimentation with macrozoobenthos and the constraints on designing effective experiments.

1.9   CONCLUSION: DECISION-MAKING FRAMEWORK

The conclusion of the book outlines a recommended decision-making process that can be followed when designing ecohydraulic experiments. It summarises the important points from each of the chapters and presents a scheme for designing such experiments that, if adhered to, will ensure that all important considerations have been taken into account.

Methods, materials and measurement

Chapter 2

Maintaining the health and behavioural integrity of plants and animals in experimental facilities

M.F. Johnson, S.P. Rice, W.E. Penning & J.T. Dijkstra

2.1 INTRODUCTION: THE IMPORTANCE OF HUSBANDRY

2.1.1 Introduction and general considerations

Once it has been decided that the experimental design requires the use of living organisms, researchers have ethical obligations to maintain organism health and to reduce organism stress where possible. This can be a challenge to some physical scientists and engineers not used to biological problems. But there are also challenges for some biologists and ecologists who may not be familiar with how environmental conditions in a laboratory flume area controlled. Physical scientists often assume that small or neurologically simple organisms, such as worms and other invertebrates, do not need to be looked after or do not have the complex responses to stimuli that are evident in larger, vertebrate animals. These assumptions are erroneous. Small organisms and plants suffer stress like any other and attention should be paid to their well-being. Similarly, these organisms interact with the physical environment in complex, subtle, but, often important ways. Consequently, organism husbandry and the conditions provided during experiments have an impact on the behaviour and physiology of small organisms and therefore on their interactions with flow and sediment that must be considered when designing them. Therefore, husbandry has two concerns: providing conditions that are not only ethical but also ensure that behaviours are analogous to wild equivalents. The following chapter emphasises the complexity of organism husbandry and the need for suitable knowledge of the organism(s) being used. It is designed to act as both a source of information for physical scientists interested in researching the hydraulic effects of organisms and for biologists and ecologists seeking a clearer understanding of the constraints presented by experimental hydraulic facilities.

2.1.2 Tolerance levels

For environmental variables like water temperature, salinity, oxygen and nutrient concentrations, tolerances exist above, or below, which organisms die. However, these tolerances are complex and difficult to generalise between species. Also, conditions that fall outside of these limits are rarely encountered in field situations. Whilst upper or lower lethal-limits exist, these are fundamentally different from threshold conditions. A threshold implies that an organism is thriving below a particular condition and stressed or dying above it. Organisms are tolerant of a wide range of conditions, but become stressed at conditions far from fatal levels. Also, stress is not just related to the magnitude of an effect, but also its duration because many organisms can survive short exposures to extreme conditions. Above the lethal limit is a "zone of resistance" where both the magnitude and length of exposure will dictate the survival or death of an organism. Animal and plant behaviour and physiology also alter in response to conditions far from those that stress the organism (Figure 2.1). Therefore, it is very important to acknowledge the sub-lethal impacts of environmental parameters on organism behaviour, morphology and physiology (see, e.g. Hill & Allanson, 1971).

Like tolerance levels, optimal conditions are difficult to generalise between species and are rarely encountered in field settings, not least because optimal temperature, hydrodynamic, sedimentary and biological variables are unlikely to occur simultaneously. Therefore, despite the obvious importance of reducing organism stress, it should also be remembered that a thriving organism under optimal conditions may not reflect the physiology or behaviour of a wild organism in natural and fluctuating field conditions. Organism husbandry would ideally aim to reflect this variability, but in practice it is extremely difficult, if not impossible, to achieve because even the best-designed laboratory experiments are highly artificial. Indeed, the whole purpose and benefit of using hydraulic facilities rather than field situations is the ability to control environmental conditions. Consequently, well-designed laboratory experiments will deviate from natural conditions by definition, but should nevertheless maintain

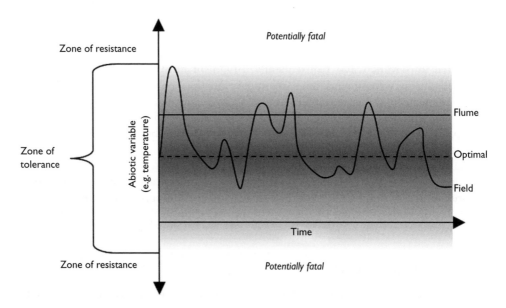

Figure 2.1 A diagrammatic illustration of the tolerance of an organism to an abiotic variable with examples of typical conditions during a flume and during a field experiment. The grey area indicates zone of tolerance whereas the area above and below, marked potentially fatal, indicates the zone of resistance which will be fatal with extended exposure.

organisms within their range of tolerance and the impact of any deviation from this should be fully acknowledged.

2.1.3   Sources of information

The importance of organism husbandry is rarely mentioned in the scientific literature and, therefore, much of the information in the following review is derived from the personal experience of researchers in the HYDRALAB IV PISCES network and their colleagues, obtained by questionnaires and interviews. As explained above, reducing organism stress should be the main aim of organism husbandry to protect the ethical and scientific integrity of the research. Information on the requirements of organisms can be found in the biological literature or by seeking expert advice from biologists and ecologists who are familiar with the species in question. Zoos and aquaria aim to keep organisms in good condition whilst recreating natural behaviours and, therefore, may also provide a pool of expert advice.

2.2   IMPORTANT CONSIDERATIONS WHEN DESIGNING EXPERIMENTS WITH ORGANISMS

2.2.1   Housing organisms prior to experiments

Organisms can either be kept in the hydraulic facility or in holding tanks and transferred prior to experiments. Both have associated challenges. If organisms are kept in holding tanks they will need to be transferred to the experimental facility and allowed to explore (if mobile) and/or acclimatise to the new conditions before starting experiments. The handling of organisms should be kept to a minimum when transferring them from and to holding tanks in order to reduce stress and/or damage. If organisms are kept in the experimental facility then the conditions prior to experiments should be similar to those during them. Similarly, holding tanks should be similar in key attributes (temperature, salinity, oxygen, etc) to the conditions that prevail during experiments.

Failure to allow sufficient acclimatisation time can lead to unrealistic behaviour or growth rates during experiments. The development and growth of plants, biofilms and some animals is also influenced by environmental conditions. An example is the architecture of biofilms, which is dependent upon the water velocity. Consequently, a biofilm acclimatised to a slow velocity will have a different architecture and, therefore, be of different morphology to those present in faster velocities (Battin et al., 2003; Chapter 5, this book).

Acclimatisation of mobile animals is characterised by increased activity as animals explore the environment. Sessile animals, such as blue mussels (*Mytilus edulis*), actively move into aggregations and their spatial distribution is related to flow and suspended food conditions (Chapter 7; van de Koppel et al., 2008). Consequently, mussels must be given time to self-organise in response to experimental conditions. The time needed for acclimatisation can vary from an hour to weeks depending on the organism being used and the study focus. For instance, measuring the shear over one mussel may require a relatively short acclimatisation period (hours), whereas a living mussel bed could require weeks of acclimatisation to achieve a stable organisation (see, e.g. van Duren *et al.*, 2006).

2.2.2   Size of the experimental area

The provision of adequate space is essential to ensure organisms behave in comparable ways to field equivalents. If animals or plants are kept in too small an area (both in terms of the plan-form area and the volume of surrounding fluid) it may stress the organism, potentially altering morphology, behaviour and/or physiology. For example, the flexible stems of some plants, including *Cabomba carloniana*, will collapse and fold in water that is too shallow. Too large an area may also have detrimental impacts, for example, wave attenuation experiments with water lilies (*Nymphaea rubra*) cannot be undertaken in deep water because floating leaves will be drowned (Penning *et al.*, 2009). Plants also require a suitable area for nutrient uptake if kept in a facility for a prolonged period. The presence of multiple individuals in small areas increases competition as resources become more limited, especially oxygen which can rapidly diminish in flumes and tanks at low flow velocities. This affects the response of organisms to environmental cues and, therefore, the impact of organisms on the hydrodynamic and sedimentary environment.

Keeping animals in a small area may also concentrate their impact. For example, many mobile animals destabilise marine sediments through their foraging and burrowing activity and the production of faecal and pseudofaecal material (Chapter 7). In the experimental area of a hydraulic facility, this effect may be concentrated as animal movement is limited; increasing the proportion of the surface with which they interact.

2.2.3   Water quality

Water chemistry is of critical importance to aquatic organisms. The first consideration should be whether the organism requires fresh or salt water. Animals requiring fresh water are likely to die in salt water and *vice versa*. Similarly, most organisms will thrive better at particular salinities. For instance, most aquatic plants cannot tolerate >10 g/L of dissolved salts (Deegan *et al.*, 2005); however, tolerance will also depend both on the type of salt (usually Na^+ or Mg^{2+}) and the duration of exposure (Llewellyn & Schaffer, 1993; Flynn *et al.*, 1995; Howard & Mendelssohn, 1999). Plant tolerance is also related to species traits and biological factors such as the life stage of the plant (Lacoul & Freedman, 2006). The salinity levels of aquatic environments are geographically variable, but on average salinity levels of 0–0.5‰ are found in freshwater environments (lakes, rivers), 0.5–30‰ in brackish water (estuaries, mangrove swamps), and 30–35‰ in seawater. Using saltwater plants in freshwater and *vice versa* will not only lead to their death, but, will also not be representative of their natural behaviour as the buoyancy of salt-water plants alters in freshwater, changing how they behave in flowing water and thus their impact on the hydrodynamic environment. Relatively few facilities can use salt water because of its corrosive properties. Therefore, experiments with saltwater may require the use of surrogates (Chapter 3). Saltwater can also be expensive to manufacture or obtain, for example, saltwater experiments at Deltares, Delft, require ocean water to be transported by tanker-ship and pumped into the flume tanks. At the Total Environment Simulator at the University of Hull, salt water is recycled from a visitor attraction (submarium) located adjacent to the facility. Subsequent disposal of saltwater may also require a permit. For freshwater organisms, standard mains-water can usually be used but it should be de-chlorinated before use.

Many studies have used water collected from, or pumped from, field sites; however, this is only suitable for small facilities and those close to a suitable source. The chemical and particulate content of the water will influence organism response. For instance, the presence of biochemicals that signal predator or prey organisms will alter their behaviour (Brönmark & Hansson, 2000; Burk & Lodge, 2002). Water may also contain bacteria, phytoplankton and other organic matter which may alter or induce feeding and foraging. Algae and bacteria could also colonise facility pumps and pipes with practical implications. Consequently, it is common to sterilise field-collected water before use.

The acidity of water is also important. For instance, aquatic plants are sensitive to acidity and can be organised into groups based on their tolerance. For instance, some species prefer softwater environments (pH < 7) whereas others are hardwater species (pH > 7). Similarly, some plants are found in brownwater environments with high concentrations of dissolved humic material (pH < 5.5) whereas plants growing in saline environments can tolerate extremely high values of alkalinity and pH (Lacoul & Freedman, 2006). The availability of carbon dioxide and the form of inorganic carbon (i.e. carbon dioxide, carbonic acid, bicarbonate, carbonate) are also related to the pH of water and exert a control on community growth and health.

Water quality can deteriorate during experiments, and pollutants, such as heavy metals, may inadvertently be added to the system from pipes and pumps. The water in hydraulic facilities may be passed through stilling tanks which often harbour viruses, bacteria and other materials that could harm or adversely affect organisms. Previous experiments may have contributed chemicals or pollutants to the water. Some laboratories have multiple facilities drawing water from a single stilling tank and, consequently, the water chemistry is dependent on experiments in all connected facilities, including any experiments running in parallel. It is for these reasons that water used for ecological experiments should ideally be changed and tanks cleaned before experiments.

2.2.4   Sediment sources and properties

The first important consideration is whether the organism lives naturally with the type of sediment being used. If it does not, then it is unlikely to behave in a way analogous to field situations. Therefore, it is important to provide plants and animals with a substrate that they colonise in nature. As with water, sediment can harbour pollutants, viruses, bacteria and various natural and man-made chemicals that could have an impact on organisms in the facility. Most natural sediments are also home to a multitude of organisms that can affect either the sediment properties or the behaviour of the target species. Biota can be removed from sediments by sieving, freezing or using biocides (Gerdol & Hughes, 1994; Lindstrom & Sandberg-Kilp, 2008). The presence of organic matter in sediments may encourage an animal to dig into the sediment or induce a foraging behaviour (Suchanek et al., 1986; Nickell et al., 1995). Plants and biofilms will also require sediments with nutrients if they are kept in a laboratory facility for an extended period (i.e. >2 weeks, discussed in section 2.2.9).

Sediment mobility under the anticipated hydraulic conditions should be avoided unless this is a specific aspect of the experiment, because mobile sediment can damage organisms by abrasion or crushing. Few plant species are able to grow on rocky substrates and coarse and cohesive sediments may offer limited rooting sites. Those that grow in cohesive sediments tend to be fine rooted plants but this sediment type is not favourable for the growth of tall plants due to its low nutrient levels (Lenssen *et al.*, 1999; Szmeja & Bazydlo, 2005). Many species can successfully anchor themselves in gravels and fine sediments (Puijalon *et al.*, 2005), but sand can also be nutrient-poor and soft and flocculated substrates may not provide sufficient anchorage (Day *et al.*, 1988; Keddy, 2000). Alternatively, plants can be kept in pots that can be inserted into a flume bed of material stable enough not to move during subsequent experiments (see e.g. Penning *et al.*, 2009; Pham *et al.*, 2011).

Studies focusing on the relationship between biota and sediment dynamics need to recreate the grain-size distribution, composition and structure of sediments. Sediments from both marine and freshwater environments have characteristic shapes which affect their stability and, consequently, the relative significance of organisms as agents of change. Some of the commercially available crushed sands or gravels may not be appropriate.

The structure of substrates is also important. For instance, fluvial substrates tend to be progressively structured by the flow, resulting in the development of coarse, imbricated armour layers. Flume studies focusing on the interaction between biota and fluvial sediments should recreate surface structure as this layer is where organisms interact (Cooper & Tait, 2009; Johnson *et al.*, 2011).

2.2.5 Water temperature

For hydraulic experiments, selection of the water temperature depends on the research design and organism being used, but it should be remembered that few hydraulic infrastructures have temperature control installed. Most plants and animals are tolerant of large temperature ranges. For instance, most temperate plant and animal species survive summer water temperatures in rivers approaching 20°C and winter temperatures as low as 0°C. Although they may survive a range of conditions, the physiology and morphology of plants and the behaviour of animals is significantly altered by temperature variations. For instance, the morphology of most temperate aquatic plants is different in winter from summer because plants become dormant with limited biomass above the sediment surface. In cold winter months, many species of aquatic fauna enter a physiological state of dormancy, termed diapause, lowering their metabolism and only increasing activity when temperature rises in the spring (Alekseev & Starobogatov, 1996).

Most species of aquatic plant have optimal rates of photosynthesis at relatively high temperatures of between 20°C and 35°C (Barko *et al.*, 1986; Santamaria & van Vierssen, 1997). Most macrophytes require ambient temperatures of at least 10°C during the growing season and most die or become dormant when temperatures drop below 3°C or rise above 45°C (Christy & Sharitz, 1980; Best & Boyd, 2003). Studies have also demonstrated that some plants grow at low temperatures, including under the cover of winter ice (Boylen & Sheldon, 1976; Olesen & Madsen, 2000).

Water temperature during experiments should be as constant as possible. Since the operation of pumps generates heat, for long or prolonged experimental runs and with powerful pumps, a water cooling system may be required to maintain consistent water temperatures.

2.2.6 Light levels

Light levels and positions affect both animals and plants. For example,careful light management is needed to limit the passage of shadows, which can disturb, stress or surprise animals and maintaining sufficient intensity and duration of light is essential for effective plant photosynthesis. Light levels should be consistent during and between experiments. Light levels also influence plant morphology (see, e.g., Liscum & Stowe-Evans, 2000; Schneider *et al.*, 2006; Barko *et al.*, 1982; Bintz & Nixon, 2001), with possible impacts on interactions between plants and hydraulics.

2.2.7 Sourcing and selecting the organisms

Individual organisms vary greatly and this can produce variability in results even under identical hydraulic conditions. In order to reduce variability, it can be beneficial to select individual organisms based on particular criteria, standardising the age or sizes used. However it should be recognised that such selection may have quantitative impacts on experimental results. For example, the size of shrimps and crabs is positively correlated to the magnitude of substrate disturbance associated with burrowing (Rowden *et al.*, 1998a; Berkenbusch & Rowden, 1999). The sex of an animal as well as any injuries or disease can also affect behaviour (see e.g., Basil & Sandeman, 2000; Koch *et al.*, 2006). Standardising plant length can be relatively simple if a plant has a homogenous structure and can be cut to length or if plants are grown from seeds in batches. However, many plants have complicated structures and branching patterns making it difficult to identify similarity in "size" and morphology among individuals.

Organisms can be sourced from the field, raised in the laboratory or bought from specialist suppliers. Plants from specialist growers have usually only experienced favourable growing conditions without competition and might, as a result, be bushier, less sturdy and have a different root/shoot ratio than wild equivalents. For animals, there may also be behavioural differences between individuals with different life histories as a result of learned behaviour in response to biological and environmental stimuli. Remarkably few examples exist of comparisons between the behaviours of laboratory-reared and field-collected species. One exception is the study of Tiselius *et al.* (1995) that compared the feeding behaviour and reproduction of copepods. They found that laboratory-reared copepods are only suitable for studying relative effects such as their activity inside and outside food layers.

Organisms collected in the field, but kept in laboratory conditions for extended periods, may undergo behavioural and physiological changes. Although understudied, questionnaires sent to leading researchers with experience of using living organisms in flumes have identified this as a potential problem. For example, one researcher noted that spinoid polychaete worms acclimated to controlled laboratory conditions, fed for extended periods relative to field equivalents because of the lack of predators. Similarly, Johnson *et al.* (2011) and Rice *et al.* (2012) noted that signal crayfish (*Pacifastacus leniusculus*) kept in captivity for 2–3 weeks became noticeably less aggressive and foraged less. Crayfish are known to form dominance hierarchies with subordinate and dominant individuals. In experiments focusing on the geomorphic impact of crayfish that utilised two crayfish, a subordinate and dominant relationship between crayfish quickly developed and was identifiable by analysing individual behaviour and activity (Figure 2.2; Rice *et al.*, 2012).

2.2.8   Selecting flow velocity and discharge

Flow velocitiy used in an experimental infrastructure must fall within the range of those experienced by the organism being researched in its natural environment to reduce unnecessary stress. The review of Lancaster & Downes (2010a) and subsequent discussion (Lamouroux *et al.*, 2010; Lancaster & Downes 2010b), illustrates the often complex response of invertebrates to abiotic conditions and warns against assuming there are simple relationships between physical variables such as flow and biota.

Many behavioural activities of aquatic animals are dependent on flow, for example, Pullen & LaBarbara (1991) showed the feeding of barnacles is dependent on flow velocity and Rice *et al.* (2008) linked the locomotion of mobile benthic animals to flow. Flow history may also influence response. For example, it has been found from streamside flumes that biofilms were thicker, had higher areal densities, lower roughness and greater sinuosity in the front edge when grown in low velocities (0.065 ±

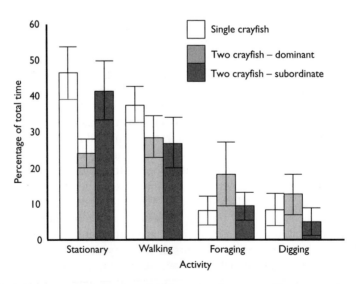

Figure 2.2 The behaviour and activity of crayfish during a six hour period broken into four discrete categories: walking, stationary, foraging, digging. White bars indicate experiments with a single crayfish and grey bars indicate experiments with two crayfish, the light grey indicating dominant individuals and dark grey indicating the activity of those that became subordinate.

0.003 m s⁻¹) in comparison to those grown in faster velocities (0.23 ± 0.005 m s⁻¹) (Battin *et al.*, 2003). Similarly, mussels and plants present in areas of high flows become short and stout in comparison to more elongated forms present in lower flows (Puijalon *et al.*, 2008; Chapter 7 this publication). Flow has also been shown to influence the attachment strength of mussels (Witman & Suchanek, 1984; Paine & Levin, 1981; Bell & Gosline, 1997; Carrington, 2002). Hunt & Schiebling (2001) found attachment strength increased from July to February when mussels were exposed to greater wave action. The history of flows may therefore influence subsequent experiments unless the individuals are carefully acclimatised.

Plants also respond to changes in flow regimes. Low velocities enhance gas exchange in comparison to still-water environments and, as a result, can lead to increased photosynthesis and growth (Smith & Walker, 1980; Crossley *et al.*, 2002). Lift and drag may break and damage plants and may even uproot them (Schutten *et al.*, 2004). Many macrophytes have a morphology that allows them to reconfigure with the flow (Vogel, 1994); for example, by flattening against the substrate or compacting leaves at high flows to reduce drag (Sand-Jensen, 2003). Some plants also align shoots in the direction of the flow and can form streamlined patches that should inform the positioning and distribution of plants in flume experiments (Idestam-Almquist & Kautsky, 1995; Sand-Jensen & Pedersen, 1999; Boeger & Poulson, 2003).

2.2.9 Nourishing organisms

Feeding animals and supplying light and nutrients to plants is essential to maintain their health. For plants, these requirements become especially important in experiments lasting longer than a week although some plants, such as some freshwater submerged plants (*Cabomba carolinaiana, Nymphaea rubra, Echinodorus grandiflorus*) can survive up to 3 weeks in a tropical flume with sufficient light but no added nutrients (WE Penning, pers.comm.). Animals may change behaviour in the presence of food, for example in mussels and polycheate worms, food causes suspension-feeding, which has implications for bed roughness (Dolmer, 2000b; Vedel *et al.*, 1994).

It is not only the quantity of food, but also the history, quality, timing and location of feeding that is important. For instance, if an animal digs for food in its natural habitat, then food should be buried in experimental facilities to ensure similar feeding behaviours. Hunger-levels are significant since a hungry animal is likely to forage more voraciously than one that is replete (Zanatell & Peckarsky, 1996). Biofilms and plants also require a nutrient-rich substrate to remain healthy, although the duration of experiments will determine the significance of this factor. Many nutrients are involved in plant growth, but carbon, nitrogen and phosphorous are the key elements required for aquatic plant life. Depending on the growth strategy, aquatic plants may obtain dissolved nutrients from either or both of the sediment and water column (Barko *et al.*, 1986; Chambers, 1987). Despite the controlling impact of nutrients and complex interactions between plant growth, morphology and nutrient limitation, it appears that most species have broad autecological amplitude with respect to nutrients. Consequently, only relatively few aquatic plants are reliable indicators of trophic conditions and nutrient availability. This is due to their inherent phenotypic plasticity as well as the ecological complexity of shallow-water environments, particularly physico–chemical variations associated with sediment and water, as well as the seasonality of inundation (Barko and Smart, 1986; Day *et al.*, 1988; Barko *et al.*, 1991). However, it has been shown that in general, plants in still-water tend to reduce total mass and number of leaves under nutrient poor conditions (Crossley *et al.*, 2002; Puijalon *et al.*, 2007).

2.2.10   Selecting population density

The density of experimental organisms influences their potential impact on hydrodynamic and sedimentary environments and increasing numbers of individuals may also increase stress levels due to competition for resources. Consequently, the impact of two organisms is unlikely to be twice that of one organism (Rice *et al.*, 2012). The scaling of animal numbers is further complicated in species that form dominance hierarchies with one dominant animal and subordinate individuals who develop significantly different behaviour. An example is crayfish, where subordinate animals become less active and aggressive in comparison to dominant individuals (Goessmann *et al.*, 2000). This has important implications when attempting to scale the impacts of mobile biota on sedimentary processes. Similar processes are also true of plants, for example, plants change morphologically or lose leaves when competition is high because of increased stress. However, the extended time it takes for these morphological changes to manifest limits their significance for most laboratory experiments (i.e., those lasting less than one week).

2.2.11   Ensuring natural behaviour

An important aspect of organism husbandry is trying to ensure that the response to, and impacts on, the hydrodynamic and sedimentary environments are analogous to those generated by wild, equivalent organisms. In order to achieve this, it may be necessary to account for other biological interactions. As discussed in sections 2.2.5 and 2.2.10, the absence of competition and predation in experimental conditions may alter the behaviour of animals and their response to abiotic variables. For example, the presence of predators undoubtedly alters the behaviour of many mobile organisms (Nyström, 2005). Other biological relationships which may be of importance to ecohydraulic studies are the presence or absence of symbiotic and epibenthic organisms. For example, there are numerous studies that have indicated both qualitatively and quantitatively that attached organisms, such as algae and barnacles, contribute significantly to the dislodgement of mussels through increasing drag (Chapter 7, this book).

The complications of biological interactions make them difficult to include in experimental design and are likely to substantially increase the experimental variance in observed phenomena due to difference in responses between organisms. Also, biological interactions may alter responses observed in studies using a single species. Numerous studies have found that the impact of species in combination is much less than the sum of the effect of those species measured in isolation (Mermillod-Blondin *et al.*, 2004a, 2005; Nogaro *et al.*, 2006; O'Reilly *et al.*, 2006). For example, Statzner & Sagnes (2009) found that barbel (*Barbus barbus*), gudgeon (*Gobio gobio*) and crayfish (*Orconectes limosus*), all of which are known to reduce surface sand levels by 10–20% in isolation, did not reduce sand levels by more than 10% when present together. Such biological interactions are probably best studied in the natural environment.

2.3 CONCLUSIONS

In general, plants and animals are tolerant of environmental variability and lethal limits are unlikely to be encountered in the wild. Indeed, many plants and animals have global ranges that demonstrate that they can thrive across a great range of environmental conditions. However, organisms are constantly responding to environmental cues in ways that impact on and alter animal behaviour and plant morphology in potentially complex and important ways. Consequently, hydraulic laboratory studies involving organisms should focus on the sub-lethal effects of environmental conditions rather than the organism's tolerances. These tolerances also interact with each other, for instance, oysters have a greater ability to survive extreme salinity at lower water temperatures and can withstand greater temperatures at optimal salinities (Heinmayer et al., 2008). The response of an organism to one environmental stressor can also completely change depending on a second stress factor. For example, Puijalon et al. (2007) showed that increasing flow velocity resulted in reduced plant size under low nutrient conditions but increased plant size under high nutrient conditions. Consequently, there are no simple answers concerning what conditions biota require in laboratory conditions or how to ensure "natural" behaviour and physiology. Nor is it possible to generalise between organisms as most have species-specific responses to environmental and biological stimuli. Tackling organism husbandry with a proper and in-depth acknowledgement of the potential complexity of organism ecology will result in better-designed experiments which are more relevant to the natural world and ethically defensible. Seeking expert ecological advice will almost always be beneficial.

Perhaps the best way of achieving good organism husbandry is to attempt to broadly replicate the habitat where the organism is found. However, it is not just the physical environment that is controlled and standardised in hydraulic facilities. A further complication that is rarely acknowledged is that the biological environment is also controlled and highly artificial in the laboratory. Ecohydraulic flume experiments tend to use only one species at a time, at controlled areal densities with standardised feeding strategies and lack of predation or competition, yet these parameters exert at least an equal control over the morphology, distribution, behaviour and physiology of organisms as environmental stimuli such as water temperature. Examples are numerous in the biological literature; an example specific to the physical impacts of organisms would be the cascade effect described by Dabourn et al. (1993). They noted that the arrival of large numbers of migratory seabirds to tidal flats in the Bay of Fundy resulted in the predation of Corophium volutator, an amphipod crustacean that destabilises sediments via bioturbation and by grazing on diatoms that stabilise sediments. Consequently, bioturbation was reduced and algal growth increased, improving the stability of the sediments (Dabourn et al., 1993). Therefore, sediment stability was governed by interactions between organisms just as much as by abiotic conditions. As stated by Lancaster & Downes (2010a, pg. 392), "it is important to realize that biological interactions do not merely add flavour to a dish largely structured by the physical environment; they are, in many situations, the main course". It is obviously not possible, or necessary, to construct an entire ecosystem within a laboratory facility, but the importance of biological interactions should be acknowledged and understood in the design stage of experiments because it will affect the spatial and temporal representativeness of experimental results in nature.

Chapter 3

Using surrogates, including scaling issues, in laboratory flumes and basins

M.F. Johnson, R.E. Thomas, J.T. Dijkstra, M. Paul, W.E. Penning & S.P. Rice

3.1 INTRODUCTION

The previous chapter outlined the challenges of looking after plants and animals hydraulic experiments. These challenges can be circumvented if inanimate surrogates can be substituted for the prototype organisms without detriment to the relevance of the data collected. Surrogates are partial replicas of biota, mimicking specific organism traits of relevance to the aims and objectives of the study. Physical surrogates range from generalised forms, such as hemispheres or plastic rods, to resin casts that precisely replicate morphology and texture. Some studies have used mechanical analogues that mimic important animal movements. Dead animal shells are regularly used as surrogates for living equivalents. There are a number of practical and research benefits to using surrogates in place of living organisms but also significant limitations. Therefore, it is important that their use is carefully considered. If large-scale processes are being studied then it may be necessary to scale down experiments so that they fit in laboratory facilities. In such studies, scaled surrogates are regularly used. These can be physical surrogates; for example, using cable ties to mimic trees, or scaled biological equivalents; for example, using small plants like alfalfa to mimic floodplain vegetation.

In employing a surrogate, the researcher is electing to study a simplified model of the real world, or prototype. Thus, Paola (2001) recognised two types of modelling approaches:

1 classical "reductionism", whereby a complete description of the prototype, in this case a real organism, is iteratively approximated until an acceptable level of simplicity is reached that available techniques are capable of realising; and
2 "synthesism", which focuses on the aspects of lower-order behaviour that dominate behaviour at a higher level, and ignores most other factors.

Neither approach is advocated herein. Nevertheless, both are justifiable and acceptable methods of approaching the same problem. A reductionist approach has been adopted by, for example, Kao & Barfield (1978), Kouwen & Li (1980), Fathi-Moghadam & Kouwen (1997), Freeman *et al.* (2000), Nepf & Vivoni (2000), Ghisalberti & Nepf (2002), Järvelä (2002, 2004), Wilson *et al.* (2003, 2005, 2006b), and Bouma *et al.* (2009c), all of whom have employed either artificial or natural flexible vegetation in field and laboratory studies. Conversely, many others have adopted a synthesist approach, using woody vegetation in its simplest form represented by wooden dowels or similar rigid structures (e.g. Petryk & Bosmajian, 1975; Pasche & Rouvé, 1985; Tsujimoto et al., 1992; Nepf, 1999; Bennett et al., 2002; Stone & Shen, 2002; Bouma et al., 2007; McBride et al., 2007; White & Nepf, 2003, 2008; James et al., 2008; Zong & Nepf, 2010) and simplified flexible structures (e.g. Bouma et al., 2005, 2009c). However, difficulties arise when adopting both approaches: the key point in Paola's descriptions of the two modelling approaches is that essential behaviours must be retained. Which behaviours are considered "essential" depends on the objectives of the study and the complexity of the problem being considered. For example, most studies have measured geometric properties of vegetation in the field and used those to inform the selection of surrogates. Some studies have also addressed the mechanical properties, however, to our knowledge only one study (Wilson et al., 2003) has also attempted to scale the mechanical properties of plants.

3.2   USING PHYSICAL SURROGATES

3.2.1   The benefits of using physical surrogates

Using physical (non-living) surrogates has a number of benefits over using living organisms, including:

- Allowing the morphology of organisms to be simplified
- Avoiding concerns for husbandry and acclimatisation
- Allowing the use of greater numbers of individuals
- Allowing complete control over the position of individuals
- Allowing detailed measurements to be made near individuals
- Allowing for spatial and temporal replication.

Each of these benefits is discussed below. There are also a number of important limitations that are discussed in section 3.2.2. Consideration of which parameters need to be mimicked and measured must inform the successful design of an experimental procedure using surrogates. It should also be noted that the benefits of using surrogates are largely practical, which in itself is not enough to justify their use. For instance, removing the need for husbandry is an advantage, but if the surrogate does not successfully mimic living equivalents this benefit may be of limited value because the results do not provide a good analogue.

3.2.1.1   Surrogates simplify the physical structure of organisms

Simplifying the physical properties of an organism creates the possibility to vary a single property while others remain constant. This allows for two types of studies: i) a reductionist approach that isolates individual variables of significance until the most important one is identified, and ii) the variation of a single variable along a range to establish a relationship that is valid for an array of natural properties. This way, it is possible to distinguish the relative importance of friction drag and form drag, or the relative importance of plant elasticity and plant buoyancy. Also, observations over a range of size or stiffness increase the applicability of the results: plants in one environment may be taller than the same plant species earlier in the season or in a different environment.

3.2.1.2   Using surrogates avoids concerns about organism husbandry and acclimatisation

A major advantage of using surrogates is that it is not necessary to consider the husbandry of the organism or its acclimatisation before experiments. Keeping an organism alive and in good condition is challenging and requires a great deal of practical management and ecological knowledge (Chapter 2). In addition, maintaining stocks of organisms may be costly, not least because of the requirement for substantial living space and the need to spend significant amounts of time feeding and cleaning organisms and their housing tanks. In addition, many physical modelling experiments have time limitations imposed by heavy use of expensive hydraulic infrastructure. This may make it difficult, or impossible, to raise or grow organisms in the facility itself unless they are very fast growing. Husbandry also impacts the physiology and behaviour of organisms, potentially generating behaviours that are not analogous to wild equivalents, with implications for experimental realism (see Chapter 2). Moreover, using live animals may simply be impossible if laboratory facilities cannot meet the minimum husbandry requirements. This is, for instance, the case when the interaction of salt-water organisms and hydrodynamics is of interest, as very few hydraulic facilities are designed to withstand salt water.

3.2.1.3   Surrogates can allow for the use of a greater number/density of organisms

One of the main benefits of surrogates is that, in comparison to live organisms, they can be cheap, expendable resources. For instance, studies of the dynamics and settlement of animal larvae in marine environments have used polystyrene balls as passive replicas of individuals (Ertman & Jumars, 1988; Grassle et al., 1992; Snelgrove et al., 1993). Larvae are difficult to raise in laboratory settings to the settlement age of approximately 3 weeks, particularly in the numbers required for multiple experimental runs. Alternatively, polystyrene balls are extremely cheap and disposable alternatives to live larvae. The associated limitation is that they only mimic passive larval dispersal, and not active swimming. Similarly, the dynamics of bivalve faecal material has been studied using couscous as a surrogate, avoiding the arduous task of collecting sufficient faecal material from field environments (De Vries, pers. comm.).

Because of the expense of growing, raising and keeping living organisms, it is unusual to take risks with, or use, large numbers of individuals. However, surrogates can be used in this way, which may promote exploration of risky or incidental research areas that nevertheless can prove fruitful. Surrogate plants have been constructed from drinking straws, flexible rubber tubes, cable ties and plastic strips that are cheaper and easier to handle than live plants. Surrogates can therefore be viewed as an important cheap alternative to living prototypes for more exploratory, early-stage research.

3.2.1.4   Surrogates allow complete control over "organism" position and preclude behavioural complications

By using surrogates, the placement of organisms in laboratory facilities or in experimental field environments can be controlled. It can be advantageous to place subject organisms in a regular grid to control non-linear interactions among individuals within a group. This is particularly the case when trying to ascertain the impacts of increased population density that could be masked by interactions between individuals due to irregular spacing. With surrogates, the proximity of individuals is easily controlled, but the same is not true for living organisms because alterations in proximity will also impact competition, with implications for the morphology and physiology of plants and the behaviour of animals (see Chapter 2, this book). Another complication of using living organisms is that even if an experiment only lasts for a relatively short period (< 12 hours), plants and animals may sicken, die, dislodge, move or breed, thus altering the biological conditions between the beginning and end of the experiment (Cardinale *et al.*, 2004).

Organisms react to environmental stimuli such as flow conditions, light and water temperature. Whilst flume experiments should in general aim to recreate natural settings, there are some situations where the aim is to study organism physiology and behaviour in unnatural environments. For instance, crayfish are known to posture in high flows, streamlining their body, and reducing drag. It is of interest to know what reduction in drag and shear is associated with this posturing. However, this can only be adequately quantified by comparing the drag and shear of postured and un-postured individuals. This is not possible with a living crayfish as there is no ethical or practical way to force them not to posture, so surrogate resin casts have been used in their place to measure drag and shear in "unnatural" situations (Johnson & Rice, unpublished).

3.2.1.5   Surrogates allow detailed measurements near the "organism"

The mobility of living organisms and their appendages can make measurements near their surfaces challenging because movement can disturb equipment or instrument sampling volumes. The movement of surrogates can be more predictable, or purposely limited, and their position can be chosen and easily changed. This makes it easier to obtain near-boundary, detailed measurements. Small-scale measurements of velocity around mobile animals can be particularly problematic because many animals will move around the facility making the identification of a single measurement location difficult. Even when stationary, many animals have moving parts including antennae, legs and tentacles that interfere with equipment, presenting further challenges to measurement. For plants, interference can be avoided or reduced by choosing surrogates with a higher rigidity than the prototype, which will lead to restricted movement under hydrodynamic forcing. However, if the interest is in flow or turbulence structure close to the plant, such a simplification will lead to unrealistic results because vegetation flexibility affects flow structures (Thompson *et al.*, 2004; Chapter 4). Also, some equipment raises ethical considerations when used in conjunction with living organisms. For example, the powerful lasers used in PIV can blind or burn organisms so that surrogates may be a better alternative.

3.2.1.6   Surrogates allow replication of experiments

The hydrodynamic environment is inherently variable and, consequently, requires replication of experimental runs to isolate treatment effects from experimental noise. The physiological and behavioural differences between individuals, or within the same individual with time, may add significantly to this variability, making it difficult to decipher signal from noise. Surrogates eliminate this individual variability, reduce the number of replicates required and therefore help minimise the number of experiments necessary to obtain consistent results. Of course, the variability associated with the physiology and behaviour of living organisms is part of their ecology, not error, so should not be dismissed when results are extrapolated and applied to the natural environment.

3.2.2   The limitations of using physical surrogates

Using surrogates introduces a number of limitations that are discussed below. The decision to use a surrogate should be made for methodological reasons, not just to reduce the challenges associated with animal husbandry. As such, it is not an alternative to ecological knowledge, which is essential in the design of a good surrogate.

3.2.2.1   Surrogates simplify the physical structure of organisms

One benefit of using a surrogate is the ability to isolate one aspect of organism morphology, for instance, mimicking the flexibility of plants (discussed in section 3.2.1.1). However, by simplifying other aspects of morphology, such as surface texture or structure, it is likely that interaction with the flow field will differ between surrogate and prototype. A good surrogate should always attempt to retain the essential behaviours relevant to a particular research question. It is difficult to assess how successful surrogates have been in mimicking living equivalents because few studies directly compare the two. A useful means of comparison is the parameterisation of important morphological attributes of living organisms. For plants, some of the attributes that should be considered in surrogate design are discussed in Frostick et al. (2011) and in Chapter 6 herein. These are:

- Stem diameter and length
- Flexibility/elasticity
- Leaf size (leaf area, length)
- Stem areal concentration (i.e. stems per unit area)
- Buoyancy/mass density.

These parameters may be combined within two non-dimensional similarity numbers (Luhar & Nepf, 2011): the Cauchy number, which is the ratio of drag and the stiffness restoring force, and the buoyancy parameter, which is the ratio of the restoring forces due to buoyancy and the stiffness. Similarity between these variables and/or these non-dimensional numbers for both living organisms and surrogate equivalents will enhance accurate reproduction of plant morphology and response to the flow. Other parameters could be added to this list, such as surface texture. A similar list does not exist for animals.

The scale of interest in the research will also influence the significance of discrepancies between surrogates and living equivalents. For instance, shell texture may change the roughness and drag on individual mussels at a scale relevant to the organism, but it may only have minimal impact on the roughness of large mussel aggregations with shells of differing heights. Similarly epiphytes on plant leaves might influence small scale studies, but may be unimportant for examining the impact of large plant stands.

3.2.2.2   *Surrogates do not replicate all organism responses*

Surrogate organisms typically mimic morphology and some aspects of flexibility in plants. However, emphasis on morphology ignores the importance of physiology and behaviour for controlling how flow affects an organism and precludes consideration of how this relationship is affected by other environmental cues. Examples include surrogate mussels that accurately model physical roughness, but do not incorporate the additional roughness associated with siphonal jets used during suspension feeding or the effects of their filtering activities on boundary layer flow development (see Chapter 7 this book).

Surrogates rarely mimic the attachment strategies of organisms even though these may be significant to some studies. For instance, O'Donnell (2008) used resin casts of mussel shells (*Mytilus californianus*) to quantify the mitigation of wave forces by mussel beds. Surrogates were firmly attached to a nested series of six concentric rings that could be detached to create circular openings within which a force transducer measured drag. However, mussels attach to substrates with byssus threads, one end of which is rubbery and the other stiff, so that the threads act as shock absorbers for the flow (Waite *et al.*, 2002). Firmly fixed surrogate mussels may not therefore replicate the wave mitigation of natural mussel assemblages. There can also be advantages to altering attachment strategies, for instance, surrogate plants are usually rootless and are tied or firmly fixed to rigid grids or bases. Consequently, they are more difficult to uproot and entrain.

3.2.2.3   *The importance of considering unexpected organism interactions*

Another limitation of using surrogates is the potential to omit an important, yet unknown or unforeseen, impact of an organism on the process being studied. For example, Butman *et al.* (1994) studied the roughness associated with living mussels (*Mytilus edulis*) over three replicate flume runs. Interestingly, during the second of the three runs, the height of mussels was 70% above that at the end of the run, creating a rougher surface. This was hypothesised to be due to mussels filtering during the second run, enticing them to orientate vertically and elevate themselves into the flow. This effect would have remained unidentified, and the realism of the experiments would, potentially, have been compromised, had surrogates been used. Minimising this limitation requires ecological knowledge of the organism being mimicked. Also, keeping research aims and objectives specific and having strong methodological reasons for using surrogates (beyond practical ease) are essential.

Table 3.1 List of plant surrogates, including the surrogate type and the parameters used for comparison.

Species	Surrogate type	Method of comparison	Comparison parameters	Purpose	Reference
Posidonia oceanica	Plastic strip	Plant parameters	Modulus of elasticity, mass density	Full scale wave measurements	Souliotis & Prinos (2009)
Posidonia oceanica	Polyethylene sheeting (Decco)	Plant parameters	Modulus of elasticity, mass density, shoot thickness, kinetic friction	Observation of turbulence properties	Folkard (2005)
Zostera marina	PVC	Plant parameters	Modulus of elasticity, mass density	Modelling of unidirectional flow structure	Dijkstra & Uittenbogaard (2010)
Zostera marina	PVC	Plant parameters and visually		Observations of turbulence properties; mixing layer	Ghisalberti & Nepf (2002)
Spartina anglica	Cable tie	Visually		Effect of growth strategies on the physical environment	Bouma et al. (2005)
Zostera noltii	PVC	Visually		Effect of growth strategies on the physical environment	Bouma et al. (2005)
Zostera noltii	Poly ribbon	Visually		Full scale wave measurements	Paul et al. (2012)
Laminaria hyperborea	Scaled plastic cast	Plant parameters (Froude scale)	Bending stiffness, modulus of elasticity	Wave attenuation	Dubi (1995); Dubi & Tørum (1995, 1997); Løvås & Tørum (2001); Wilson et al. (2003)
Cabomba caroliniana and Egeria densa	Two flexible artificial representatives of these macrophytes and a rigid cylinder				Cooper et al. (2007)
Phragmites australis	Plastic ornamental plants			Observation of flow resistance by flexible vegetation	Vionnet et al. (2004)
Zostera japonica Spartina anglica	Nylon ribbon Semi-rigid Nylon tube and Polythene sheet	Plant parameter	Modulus of elasticity, plant geometry	Effect of seagrass on fauna Measuring suspended sediment dynamics	Lee et al. (2001) Graham et al. (2007)
"Grass"	Display grass and AstroTurf	Visual from photographs	Blade height and density	Flow resistance	Wilson et al. (2005)
Poplar, Willow, Sedge	Artificial, plastic plants	Plant height and area under no flow, leaf number and area	The lever arm of plants under drag	Drag force	Schoneboom et al. (2008, 2010)

3.2.3 Mimicking plants

Plants have been mimicked in order to investigate their effect on the hydrodynamic environment, under both unidirectional flow and wave motion. The highest degree of simplification uses solid cylindrical (Nepf *et al.*, 1997; Tanino & Nepf, 2008) or strip like structures (Augustin *et al.*, 2009) and, therefore, neglects plant motion due to flexibility. However, the importance of vegetation stiffness with respect to flow and wave attenuation has been recognised (Koehl, 1996; Bouma *et al.*, 2005) and incorporated in many studies (Table 3.1).

3.2.3.1 Flexible full-scale surrogates

Various methods have been used to develop surrogates with stiffnesses similar to their real plant prototypes. For example, Bouma *et al.* (2005) used plastic strips in unidirectional flow. In addition, they decided upon the surrogate by visually comparing the bending behaviour with real equivalents. This was deemed a suitable technique given that the focus of the study was to identify differences between a salt marsh and seagrass plant species with very different bending characteristics (*Spartina anglica* vs. *Zostera noltii*). Paul *et al.* (2012) also developed surrogates of *Zostera noltii* by visual comparison, living plants were placed in rows with different surrogate materials included in each row and the degree of deflection was compared through the transparent flume wall (Figure 3.1). Their observations showed that poly ribbon had greater similarity to living *Zostera noltii* than the plastic strips used by Bouma *et al.* (2005). The tests of Paul *et al.* (2012) also included lengths of wool that matched well the leaf dimensions of *Zostera noltii* but proved to be non-buoyant and were, therefore, unsuitable to represent buoyant vegetation. Both sets of authors developed

Figure 3.1 Comparison of different surrogate materials with live Zostera noltii shoots under unidirectional flow (M. Paul).

their surrogates under unidirectional flows, but used the surrogates in experiments with wave conditions. Consequently, both studies assume unidirectional flows and orbital wave motion influence the surrogate in the same way, an assumption that is, as yet, untested.

Ghisalberti & Nepf (2002) developed a surrogate for the seagrass *Zostera marina* by comparing physical plant properties, including the modulus of elasticity (E) and the density of leaf tissue, to possible surrogates. The living plants showed natural variation in these properties so a surrogate was chosen with material properties that were within the range of the prototype whilst maximising variation. The resultant surrogates have been used in several subsequent studies in unidirectional flow (e.g. Ghisalberti & Nepf, 2006) and under wave motion (e.g. Luhar *et al.*, 2010). Other studies have used tissue density (ρ_t) and modulus of elasticity to find surrogate materials that represent seagrass species (Fonseca & Koehl, 2006; Manca, 2010), but similarity in these parameters did not automatically lead to similar behaviour under hydrodynamic forcing, which suggests that other parameters may be important. Wilson *et al.* (2005) used AstroTurf and plastic grass as surrogates for real grass in a study of flow resistance to overland flow. The real grass had average blade heights of 20 and 70 mm in comparison to display grass and AstroTurf where this was 6 and 16 mm, respectively. The velocity at half flow depth was 25% less over AstroTurf than real grass. It was hypothesised that the differences in the hydraulic behaviour of real and artificial grasses arose because rotational movements in real grasses were not mimicked by the surrogates (Wilson *et al.*, 2005).

Schoneboom *et al.* (2008, 2010) studied the drag force acting on both natural and artificial poplar and willow and introduced a new parameter, the lever arm, to quantify plant dynamics. The lever arm is the perpendicular distance between the line of action of a force and the point on which an object pivots. Schoneboom *et al.* (2008) found that the force-velocity and lever arm-velocity relationships were almost identical in surrogates and real poplars up to a velocity of 1 m s^{-1}. Above this, the lever-arm of natural poplar suddenly decreased, diverging substantially from the surrogate. Schoneboom *et al.* (2008) associated this decrease with an exceedance of the maximum load on the stem, resulting in the flexural rigidity of the poplar changing significantly. Larger differences were measured between artificial and real willows and, consequently, surrogates were not considered appropriate. They conclude that using drag force measurements with a consideration of lever arm is a good way of validating surrogate design against natural equivalents.

3.2.3.2 Scaled physical surrogates of plants

Surrogates of small plants can often be used without scaling. For large plants, such as mangroves, kelp or floodplain trees and shrubs, scaling is necessary to carry out experiments in laboratory facilities. For example, scaled surrogates have been developed for the root system of mangroves (Husrin & Oumeraci, 2009) and in this case, because mangrove roots and stems are rigid, scaling only considered shape and did not need to take flexural stiffness or elasticity into account. Scaled models of the kelp *Laminaria hyperborea* were created by Dubi (1995) from liquid plastic. The surrogates successfully represented the shape of plants at 1:10 scale, but the stiffness was identical to the living plant and, consequently, was not scaled (Løvås & Tørum, 2001).

When scaling surrogates, scale factors need to be considered. For example, if the linear scale relating the stem diameter between a surrogate and the living equivalent is $1/X$, then the ratio of areas between the model and its prototype is simply $1/X^2$. Identifying other scale factors requires the selection of a similarity principle. Most hydraulic studies adopt Froude similarity, ensuring that the ratio between the rate of fluid flow and the rate of gravity wave propagation is constant. If Froude similarity is adopted, some pertinent parameters scale as shown in Table 3.2. For example, the velocity in the model must be scaled with a factor of $X^{-1/2}$ relative to the prototype, time will be similarly scaled (i.e. process rates will speed up by a factor of $X^{1/2}$ relative to the prototype), the applied (drag) force will be scaled by a factor of X^{-3} relative to the prototype and to ensure appropriate behaviour, the flexural rigidity

Table 3.2 Derivation of scaling factors for pertinent parameters assuming Froude and Reynolds similarity. Estimation of the force, bending stiffness and flexural rigidity scales assumes that the classical drag equation $(F = \frac{1}{2}C_D\rho_w Au^2)$ holds for all flow stages and that both the drag coefficient, C_D, and the density of water, ρ_w, are unchanged between model (m) and prototype (p).

Parameter	Froude similarity	Reynolds similarity
Velocity	$$F_r = \frac{u_m}{\sqrt{gh_m}} = \frac{u_p}{\sqrt{gh_p}}$$ $$u_m^2\,gh_p = u_p^2\,gh_m$$ $$\frac{u_m}{u_p} = \sqrt{\frac{h_m}{h_p}} = \frac{1}{\sqrt{X}}$$	$$R_e = \frac{u_m d_m}{v} = \frac{u_p d_p}{v}$$ $$u_m d_m = u_p d_p$$ $$\frac{u_m}{u_p} = \frac{d_p}{d_m} = X$$
Time	$$t_m = \frac{h_m}{u_m}, t_p = \frac{h_p}{u_p}$$ $$\frac{t_m}{t_p} = \frac{h_m u_p}{h_p u_m} = \frac{1}{X}\sqrt{X} = \frac{1}{\sqrt{X}}$$	$$t_m = \frac{h_m}{u_m}, t_p = \frac{h_p}{u_p}$$ $$\frac{t_m}{t_p} = \frac{h_m u_p}{h_p u_m} = \frac{1}{X}X = 1$$
Force (drag force equation $F = \frac{1}{2}C_D\rho Au^2$, where C_D = drag coefficient, ρ_w = density of water, A = frontal/projected area, u = mean undisturbed velocity)	$$F_m = \frac{1}{2}C_D\rho_w A_m u_m^2,\ F_p = \frac{1}{2}C_D\rho_w A_p u_p^2$$ $$\frac{F_m}{F_p} = \frac{A_m u_m^2}{A_p u_p^2} = \frac{1}{X^2}\frac{1}{X} = \frac{1}{X^3}$$	$$\frac{F_m}{F_p} = \frac{A_m u_m^2}{A_p u_p^2} = \frac{1}{X^2}X^2 = 1$$
Bending stiffness $(M = F/\omega$, where ω = deflection of object from position at rest caused by the force F)	$$M_m = \frac{F_m}{\omega_m} = \frac{\frac{1}{2}C_D\rho_w A_m u_m^2}{\omega_m},$$ $$M_p = \frac{F_p}{\omega_p} = \frac{\frac{1}{2}C_D\rho_w A_p u_p^2}{\omega_p}$$ $$\frac{M_m}{M_p} = \frac{A_m u_m^2 \omega_p}{A_p u_p^2 \omega_m} = \frac{1}{X^2}\frac{1}{X}X = \frac{1}{X^2}$$	$$\frac{M_m}{M_p} = \frac{A_m u_m^2 \omega_p}{A_p u_p^2 \omega_m} = \frac{1}{X^2}X^2 X = X$$

(Continued)

Table 3.2 (Continued).

Parameter	Froude similarity	Reynolds similarity
Flexural rigidity ($J = EI$, where E = Young's modulus of elasticity = Stress/strain = $(F/A)/(\Delta L/L)$, ΔL = amount by which the length, L, of an object changes caused by the force F and I = moment of inertia, which is approximated for a cylinder by $\frac{\pi d^4}{64}$	$J_m = E_m I_m = \dfrac{F_m L_m I_m}{A_m \Delta L_m}$ $= \dfrac{1}{2} C_D \rho u_m^2 \dfrac{L_m}{\Delta L_m} \dfrac{\pi d_m^4}{64}$, $J_p = E_p I_p = \dfrac{1}{2} C_D \rho_w u_p^2 \dfrac{L_p}{\Delta L_p} \dfrac{\pi d_p^4}{64}$ $\dfrac{J_m}{J_p} = \dfrac{u_m^2 L_m \Delta L_p d_m^4}{u_p^2 L_p \Delta L_m d_p^4}$ $= \dfrac{1}{X} \dfrac{1}{X} \times \dfrac{1}{X^4} = \dfrac{1}{X^5}$	$\dfrac{J_m}{J_p} = \dfrac{u_m^2 L_m \Delta L_p d_m^4}{u_p^2 L_p \Delta L_m d_p^4}$ $= X^2 \dfrac{1}{X} \times \dfrac{1}{X^4} = \dfrac{1}{X^2}$
Density of material to ensure correctly scaled fundamental bending frequency $f = \dfrac{\pi}{2L^2}\sqrt{\dfrac{EI}{\sigma A}}$, where ρ = density of material)	$f_m = \dfrac{\pi}{2L_m^2}\sqrt{\dfrac{E_m I_m}{\sigma_m A_m}}$, $f_p = \dfrac{\pi}{2L_p^2}\sqrt{\dfrac{E_p I_p}{\sigma_p A_p}}$ $\dfrac{f_m}{f_p} = \sqrt{X}$ $= \dfrac{\pi}{2L_m^2}\sqrt{\dfrac{E_m I_m}{\sigma_m A_m}} \Big/ \dfrac{\pi}{2L_p^2}\sqrt{\dfrac{E_p I_p}{\sigma_p A_p}}$ $\sqrt{X} = \dfrac{L_p^2}{L_m^2}\sqrt{\dfrac{E_m I_m \sigma_p A_p}{E_p I_p \sigma_m A_m}} = X^2\sqrt{\dfrac{1}{X^5}\times X^2 \dfrac{\sigma_p}{\sigma_m}}$ $X^{1/2}X^{-2}X^{3/2} = 1 = \sqrt{\dfrac{\sigma_p}{\sigma_m}}$ $\dfrac{\sigma_m}{\sigma_p} = 1$	$f_m = \dfrac{\pi}{2L_m^2}\sqrt{\dfrac{E_m I_m}{\sigma_m A_m}}$, $f_p = \dfrac{\pi}{2L_p^2}\sqrt{\dfrac{E_p I_p}{\sigma_p A_p}}$ $\dfrac{f_m}{f_p} = 1$ $= \dfrac{\pi}{2L_m^2}\sqrt{\dfrac{E_m I_m}{\sigma_m A_m}} \Big/ \dfrac{\pi}{2L_p^2}\sqrt{\dfrac{E_p I_p}{\sigma_p A_p}}$ $\dfrac{L_p^2}{L_m^2} = \sqrt{\dfrac{E_p I_p \sigma_m A_m}{E_m I_m \sigma_p A_p}}$ $X^2 = \sqrt{X^2 \dfrac{1}{X^2}\dfrac{\sigma_m}{\sigma_p}} = \sqrt{\dfrac{\sigma_m}{\sigma_p}}$ $\dfrac{\sigma_m}{\sigma_p} = X^4$

should be scaled by a factor of X^{-5} relative to the prototype. Therefore, if a model is a 1/10 scale replica of its prototype, the applied force will be 1/1000 that observed in the prototype system and to ensure reasonable behaviour, the flexural rigidity should be 1/100000 that observed in the prototype. This is not as significant a problem as it may first appear because a significant proportion of this factor arises from the moment of inertia, which is a geometric quantity that scales as X^{-4}, or 1/10000 for this example. Thus, it is only necessary to scale Young's modulus of elasticity by a factor of 1/10. Appendix 1 presents typical Young's moduli of some common materials and riparian plant species.

Alternatively, if it is desirable to maintain the scale of turbulent eddies, stem Reynolds number may be used as the basis of similarity. In this case, the velocity in the model must be scaled with a factor of x relative to the prototype, the applied (drag) force will be identical to that observed in the prototype and to ensure appropriate behaviour, the flexural rigidity should be scaled by a factor of X^{-2}. If a model is a 1/10 scale replica, the velocity should be 10 times, the flexural rigidity should be 1/100, and, as the moment of inertia scales as X^{-4} or 1/10000 in this example, Young's modulus of the model should be 100 times that of the prototype. In addition, to ensure that stems oscillate with the correctly scaled fundamental bending frequency, the density of the material should be 10000 times the prototype, which is extremely difficult to achieve while simultaneously fulfilling the other constraints.

3.2.4 How previous experiments have mimicked animals

Animals and their constructions provide obstacles to flow that have long been studied using standardised shapes such as hemispheres and tubes (Eckman, 1983; Eckman & Nowell, 1984; Davidson *et al.*, 1995; Friedrichs *et al.*, 2009). A hemisphere can represent a stationary crustacean, an individual shell, or a mound of excavated bed material. However, the generality of using such shapes means no organisms are accurately mimicked in detail. Consequently, this approach is excellent for interpreting general trends, but cannot be used to mimic detailed, species-specific or small-scale hydrodynamic impacts around biota.

Species-specific surrogates of animals have been used, including artificial mimics of mussel, gastropod and barnacle aggregations (Pullen & LaBarbera, 1991; Crimaldi *et al.*, 2002; Moulin *et al.*, 2007; O'Donnell, 2008; Folkard & Gascoigne, 2009). In general, these have been constructed from dead shells, resin casts, multiple aggregated tubes and clay-modelled blocks (Table 3.3). Whilst these surrogates identify the general hydrodynamic processes operating over organisms, they do not model animal movement, which can be significant, even for sessile animals. For instance, Pullen & LaBarbera (1991) used plasticene-filled plastic tubes as surrogates of barnacles and attached netting to the tops of tubes to mimic suspension-feeding appendages. While this was appropriate for passive feeding, it did not mimic the active flicking of fans which can have a significant hydrodynamic effect (Thomason *et al.*, 1998; Chapter 7 this book). Animal surrogates have also been used to investigate the feeding flows generated by bivalves. Monismith *et al.* (1990) constructed physical models of siphons using tubes from which water was jetted at velocities analogous to the ejections of the clam *Tapes japonica*. While this study identified potential impacts of jets on turbulent mixing and roughness, it did not account for variations in siphonal use by bivalves. For instance, excurrent siphons, and to a lesser degree incurrent siphons, contract and expand during pumping (Troost *et al.*, 2009). In addition, siphons can be moved vertically and horizontally to reduce refiltration (O'Riordan *et al.*, 1995; Troost *et al.*, 2009 and Chapter 7 this book).

Surrogates of mobile animals are rare, but do exist. Lim & DeMont (2009) constructed a robotic lobster to study the hydrodynamics associated with the beating of pleopods (paddles under the tail) in order to assess whether this aided locomotion (jet-walking). The tail of a dead crayfish, including the underlying pleopods,

Table 3.3 Studies using surrogate animals and the type of surrogate used.

Animal	Surrogate type	Attachment	Purpose	Arrangement	Ref.
Mussels/ Oysters	Dead shells	Placed in resin	Impact on near-bed flow		Folkard & Gascoigne (2009)
	Resin casts	Adhesive	Wave dampening	Nested, concentric rings	O'Donnell (2008)
Clams	Dead shells	Embedded in clay	Impact on near-bed flow		Crimaldi et al. (2002)
Bivalve siphons	Tubes (variable diameter)	Embedded, exposed 0–10 mm above surface	Biologically generated flows and roughness	Regular grids or individuals	Monismith et al. (1990); Crimaldi et al. (2007)
Gastropod (Crepidula fornicata)	Resin casts	Adhered to plane bed	Impact on near-bed flows	Regular grids, natural densities	Moulin et al. (2007)
Anemone	Clay, plaster, fabric, aluminium, plastic foam		Drag measurements	Individuals	Koehl (1977)
Barnacles	Plastic cylinders (9.7 mm diameter) with netting fans	Firmly fixed	Flow and filter-feeding	Linear rows of different heights	Pullen & LaBarbera (1991)
	Resin casts of colonies	Firmly fixed	Hydrodynamics	Natural representation	Thomason et al. (1998)
Polychaete lawn	Tubes (5.4 mm diameter)	Embedded in sand (35 mm protrusion)	Impact on near-bed flow	Regular pattern	Friedrichs et al. (2000)

was attached to a frame and wire was threaded into the pleopods. Each wire thread was connected to servomotors which moved the wire in a way that mimicked the motion being studied. The advantage of using a surrogate rather than a living organism in that study was that the timing of pleopod beating could be controlled, generating a regular, predictable pattern that could be started and stopped on cue (Lim & DeMont, 2009).

3.2.5   The questions that should be asked before using physical surrogates

Using physical surrogates in place of living organisms can be an extremely useful tool in identifying the interactions of organisms and flow. However, as stated above, a number of limitations mean the use of surrogates needs to be carefully considered. The above review has illustrated some of these considerations which are listed below. These questions should be asked before designing experiments using surrogates:

- Why use a surrogate – is it possible to use living equivalents?
- Would it be possible to use a living equivalent in a different facility or with different equipment?
- What is the scale of the study, and what scale of results are of interest? Does this impact the design of surrogates?
- What organism characteristics need to be mimicked in order to successfully research the aims and objectives?
- Does the surrogate suitably mimic these characteristics?
- What will the surrogate be constructed from and is that material suitable?
- Is there a danger that unknown behaviours or physiological effects might compromise results? Can this problem be removed by redesigning the methodology?

3.3   USING ALTERNATIVE ORGANISMS AS SURROGATES

3.3.1   The benefits of using other, living organisms as surrogates

The use of biological surrogates is relatively limited but has been successfully adopted in studies of river floodplains. Vegetation influences physical processes on floodplains, affecting fluvial channel behaviour, planform, morphology and sediment stability (Tal & Paola, 2007, 2010). It is not possible to study such processes in experimental flumes at a 1:1 scale. Instead, they have to be scaled down. Scaling such environmental factors, including sediments and flows is discussed in detail in Frostick *et al.* (2011) and elsewhere (Nowell & Jumars, 1987; Peakall *et al.*, 1996). This approach to scaling biota has not been widely considered. However, vegetation has been scaled by using fast-growing, small plants such as alfalfa to represent large-scale plants (Gran & Paola, 2001; Tal & Paola 2007, 2010; Van de Lageweg *et al.*, 2010).

3.3.2 The limitations on using biological surrogates

The limitations of biological, scaled surrogates are similar to those for physical surrogates, for example, using a surrogate may fail to suitably replicate responses to abiotic variables. The main limitations of their use include:

- They tend to be selected based on their practical benefits (fast growth, small size), rather than their accuracy in representing larger equivalents
- They pose the same problems as using other living species in that some knowledge of their husbandry is required (see Chapter 2 this book).

3.3.3 Commonly-used biological surrogates

Alfalfa (*Medicago sativa*) is a species of plant that has been used in past experiments as a surrogate for larger vegetation. This is because of its small size, rapid growth and the success of experiments that pioneered its use (Gran & Paola, 2001). These authors studied the impact of vegetation on river planform in a flat-bed flume (16 m long, 2 m wide) using alfalfa and non-cohesive sand ($D_{50} = 0.5$ mm). Alfalfa seeds were distributed uniformly over an experimental area. Some seeds landed on exposed areas and germinated, whereas others landed in the flow and were washed downstream and deposited in marginal areas. This replicated the dispersal of seeds by many riparian trees such as willow and cottonwood. As riparian vegetation density increased, channel form and dynamics were altered due to the increased bank strength. Tal & Paola (2007, 2010), building on the results of Gran & Paola (2001) and following the same methodology, found the presence of alfalfa resulted in a shift from braided channels to single-threaded, meandering channels with vegetated floodplains. In these experiments, alfalfa seeds were soaked for 48–72 hours and then air dried for 6–12 hours before deployment in experiments. After 10–14 days the alfalfa sprouts had a single stem roughly 30 mm in length and 1 mm in diameter with two to four small leaves on top. The root system penetrated 30 mm deep and consisted of a taproot with smaller branching rootlets (Gran & Paola, 2001). Therefore, the morphology of alfalfa is most similar to trees with a solitary trunk and high branches. Pine trees fit this description and are taproot dominated; however, the applicability of results to floodplains that are not dominated by dense pines is unknown. In fact, most temperate floodplains are characterised by plate-root-dominated networks (e.g. shallow rooting riparian willow species) or heartroot networks (e.g. river birch and sycamore trees), both very different from the morphology replicated by alfalfa. A 1 mm diameter alfalfa sprout in these experiments scaled to an 80 mm diameter tree based on the relationship between the sediment in the experiments and at the field site (Sunwapta River, Alberta, Canada). In addition to replicating the gross morphology of trees, the cohesion provided by alfalfa roots (0 to 11.8 kPa) is similar to the magnitude of root-reinforcement expected under field conditions (Pollen & Simon, 2006). Braudrick et al. (2009) also used alfalfa in flumes to mimic floodplain vegetation and provide bank strength. They found alfalfa also increased flow resistance, promoting fine sediment deposition and, consequently, retarding chute-cutoff and causing vertical accretion on bars and exposed islands (Braudrick et al., 2009).

Van de Lageweg *et al.* (2010) assessed the potential of four plant species to mimic large floodplain vegetation. The species were alfalfa (*Medicago sativa*), garden cress pepperweed (*Lepidium sativum*), garden rocket (*Eruca sativa*) and thale cress (*Arabidopsis thaliana*). All the plants germinated and grew quickly on fully saturated soils; however, garden cress pepperweed and garden rocket germinated faster, yielding taller plants that had more significant root systems than both alfalfa and thale. Garden cress pepperweed and garden rocket had a single deep penetrating main root, but only the root of the latter had sideways branching rootlets providing better anchorage. Use of a grow lamp resulted in plants developing a relatively small stem height, but with thicker stems and leaves in comparison to control plants without additional lighting (van de Lageweg *et al.*, 2010). As a result of these experiments, these authors undertook a series of experiments using garden rocket grown under a grow lamp as a surrogate and found that bank stability increased with increased stem areal density. However, they also noted that when vegetation was sparse, erosion could be increased locally due to the turbulence generated around individual plants.

3.3.4 Considerations before using a biological surrogate

Based on the above review, a number of factors need to be considered before using scaled biological surrogates. These include:

- What parameters should be scaled: leaf area; stem diameter; flexibility; height; width; root mass?
- Not all dimensions can be exactly scaled, hence no biological surrogate will exactly mimic the effect of a larger plant – does this impact the aims of the study?
- The surrogate may broadly mimic the morphology of a larger plant, but does it behave in a similar way, for instance, does the plant respond to the flow in the same way?
- What areal density of surrogates will be used and how does this compare to natural equivalents?
- Is it possible (or necessary) to scale biological interactions, such as plant competition?
- Will the plants be grown in the flume channel or in a tank? Will transfer to the flume cause disturbance?
- Will growing conditions affect plant morphology? (see Chapter 2 this book).

3.4 CONCLUSIONS

Using surrogates that mimic organisms is a useful way of investigating the complex interaction between the hydraulic environment and biota because using surrogates overcomes many of the challenges associated with organism husbandry and allows more flexibility in the use of measurement techniques. The use of surrogates is, therefore increasingly popular. Ideally, the decision to use a surrogate, the choice of materials and the surrogate's design follow a similar procedure to any other project; i) identify knowledge gaps and key variables, ii) identify a means to fill those gaps (in

this case, using surrogates), iii) measure key variables in the field/prototype and identify a suitable scaling technique if relevant, iv) perform the experiment, and v) validate
the experiment. However, published accounts often lack a clear description of how
surrogates were selected, obtained or constructed. There has also been little attempt to
validate results obtained with surrogates by using living equivalents. In addition, there
is lack of understanding of which parameters of living organisms should be replicated
and how parameters should be scaled to mimic larger equivalents. Often, only visual
comparison is undertaken, which is likely to be unsuitable in some cases. Therefore,
the ability of surrogates to successfully replicate important characteristics of living
organisms, and, more generally, the development of methods for making comparisons
between surrogates and their prototypes, have not been adequately considered.

3.5   APPENDIX

Typical values of Young's modulus of elasticity for some common materials, assem-
bled from various sources.

Material	Manufacturer's specified Young's modulus of elasticity $(10^9 \ N \ m^{-2})$
Acrylonitrile-Butadiene-Styrene plastics	2.30
Acrylic	2.09–3.20
Aluminium	69.0
Antimony	77.9
Beryllium	290
Bismuth	31.7
Bone	9.00
Brasses	96.0
Bronzes	113
Cadmium	31.7
Chromium	248
Cobalt	207
Concrete, High Strength (compression)	30.0
Copper	117
Diamond	1130
Fibreglass	41.4
Glass	70.0
Gold	74.5
Iridium	517
Iron	197
Lead	13.8
Magnesium	45.0
Manganese	159
Molybdenum	276
Nickel	214
Niobium (Columbium)	103
Nylon	3.00

(Continued)

(*Continued*).

Material	Manufacturer's specified Young's modulus of elasticity $(10^9\ N\ m^{-2})$
Oak Wood (along grain)	11.0
Osmium	552
Platinum	147
Plutonium	97
Polycarbonate	2.22–2.60
Polyethylene HDPE	0.80
Polyethylene Terephthalate PET	2.35
Polyimide	2.50
Polypropylene	1.75
Polystyrene	3.25
Rhodium	290
Rubber	0.055
Selenium	57.9
Silicon	110
Silicon Carbide	450
Silver	72.4
Steel, Structural (Conforming to ASTM International standard A36)	200
Tantalum	186
Thorium	58.6
Titanium	110
Titanium Alloy	113
Tungsten	405
Tungsten Carbide	550
Uranium	166
Vanadium	131
Wood dowel (Small)	17.9
Wood dowel (Large)	2.86
Wrought Iron	200
Zinc	82.7

Listing of typical values of Young's modulus of elasticity for some common riparian plants. From Freeman *et al.* (2000) unless otherwise stated.

Common name	Scientific name	Young's modulus of elasticity $(10^9\ N\ m^{-2})$
Alder	*Alnus Incana*	1.70
White Ash (bark) (Niklas 1999)	*Fraxinus americana*	0.52 ± 0.053*
White Ash (wood) (Niklas 1999)	*Fraxinus americana*	1.05 ± 0.090*
Yellow Twig Dogwood (sample 1)	*Cornus Stolonifera Flaviramea*	0.32
Yellow Twig Dogwood (sample 2)	*Cornus Stolonifera Flaviramea*	2.99

(Continued)

(Continued).

Common name	Scientific name	Young's modulus of elasticity (10^9 N m^{-2})
Red Twig Dogwood	Cornus Sericea	1.02
Berried Elderberry	Sambucus Racemosa	0.053
Blue Elderberry	Sambucus Canadensis	0.026
Valley Elderberry	Sambucus Mexicana	1.65
Purpleleaf Euonymus	Euonymus Fortunei Colorata	0.41
Service Berry	Amelanchier	4.76
Mulefat	Baccharis Glutinosa	0.60
Norway Maple	Acer Platenoides	1.91
Sugar Maple (bark) (Niklas 1999)	Acer saccharum	0.34 ± 0.008*
Sugar Maple (wood) (Niklas 1999)	Acer saccharum	0.83 ± 0.032*
English Oak (bark) (Niklas 1999)	Quercus robur	0.54 ± 0.049*
English Oak (wood) (Niklas 1999)	Quercus robur	1.07 ± 0.090*
Common Privet	Ligustrum Vulgare	0.39
Common reed (Ostendorp 1995)	Phragmites australis	19.3 ± 3.0
Common reed (Boar et al. 1999)	Phragmites australis	5.92 ± 0.50
Wild Rose	Rosa spp.	13.0
Salt Cedar	Tamarix spp.	1.31
Western Sand Cherry	Prunis Besseyi	2.88
Staghorn Sumac	Rhus Typhina	0.51
Sycamore	Platenus Acer Ifolia	2.75
Arctic Blue Willow	Salix Purpurea Nana	0.12
Black Willow	Salix Nigra	0.15
French Pink Pussywillow	Salix Caprea Pendula	0.11
Lemmon's Willow	Salix Lemmonii	8.60
Mountain Willow	Salix Monticola	0.34
Pacific Willow	Salix lasiandra	9.90
Red Willow	Salix spp.	0.45
Sand Bar Willow	Salix exigua	8.62

* for each species, standard deviations were computed using averages presented by Niklas (1999) for plant ages ranging from 0.5–11 years and thus do not represent the full range of variability.

Chapter 4

Flow measurement around organisms and surrogates

R.E. Thomas, S.J. McLelland, O. Eiff & D.R. Parsons

4.1 INTRODUCTION

This chapter reviews methods of measuring turbulent fluid flowing through and around flora and fauna. It begins with a discussion of the relevant aspects of the nature of turbulent flows and highlights difficulties associated with their measurement. Detailed discussions of turbulent flows can be found in many textbooks, including Tennekes & Lumley (1972), Hinze (1975), Nezu & Nakagawa (1993), Pope (2000) and Davidson (2004) and are not repeated here. Here, the focus is on generic issues that affect the successful measurement of turbulent flows, irrespective of the technique adopted and the crucial issue of selecting an appropriate measurement duration. There is a summary of relevant techniques with a detailed review of how each one has been used with biota or surrogates, including past deployments, and an overview of relevant strengths and weaknesses. Finally some suggestions about standardised reporting of instrument deployment and data processing are presented, which if adopted, would provide adequate grounds for critical evaluation of reported data and promote best practice.

4.1.1 Time and space scales of turbulent flows: Implications for measurement

Turbulent fluid flows are highly irregular in space and time, and thus encourage the use of statistical rather than deterministic methods to describe them. The challenge that faces experimentalists is to quantify spatio-temporal variability, such that the resulting measured statistics adequately represent the actual statistics of the flow. Turbulent flows are comprised of velocity fluctuations so that they are commonly quantified by decomposing the flow into a time-average (or Reynolds-average, after Reynolds, 1895) and fluctuations about that average. i.e.

$$u_i = \overline{u_i} + u_i' \tag{4.1}$$

$$\overline{u_i} = \lim_{\Delta t \to \infty} \frac{1}{\Delta t} \int_{t_0}^{t_0 + \Delta t} u_i(t)\, dt \tag{4.2}$$

where i ranges from 1 to 3, corresponding to three orthogonal velocity components $u_i = \overline{u_i} + u'$ (streamwise), $v = \overline{v} + v'$ (vertical), and $w = \overline{w} + w'$ (cross-stream), overbars represent time-averages over the averaging interval Δt, primes represent fluctuations about that average, t_0 is the time at the start of the time-averaging period, d is the differential operator, and the velocity has been written in (4.2) as $u_i(t)$ to explicitly indicate that it varies with time. Note that by definition, $\overline{u_i'} \equiv 0$; the spread of the distribution of u_i' is commonly quantified using the root mean square fluctuation, $\sqrt{\overline{u_i'^2}}$.

The Reynolds decomposition is only really useful or even meaningful if the mean flow is constant (or nearly constant) over time. Given a steady mean velocity (referred to as a stationary flow by Tennekes & Lumley, 1972), periods of relatively accelerated flow must be balanced by periods of relatively decelerated flow. The term "period" can be considered to refer to a period of time or a region of space within the fluid. Thus, although turbulent flows are irregular, they also exhibit periodicity. To describe this periodicity, the extent to which velocities $u_i(x_i, t)$ measured at the location x_i ($i = 1$ to 3) at time t are related to the velocities measured at the same location at time $t \pm \Delta t$ is estimated using the temporal autocorrelation coefficient $\alpha(\Delta t)$:

$$\alpha(\Delta t) = \frac{\overline{u_i'(x_i, t)u_i'(x_i, t \pm \Delta t)}}{\overline{u_i'^2}} \tag{4.3}$$

The smallest timescale of the flow (λ) can be estimated using the second derivative of $\alpha(\Delta t)$ at a lag, Δt, of zero (Tennekes & Lumley, 1972):

$$\frac{d^2\alpha(0)}{d\Delta t^2} \equiv -\frac{2}{\lambda^2} \tag{4.4}$$

After noting that $d^2\overline{u_i'^2}/dt^2 = 0$ and performing some algebraic manipulation, Tennekes & Lumley (1972) obtain:

$$\overline{\left(\frac{du_i'}{dt}\right)^2} = \frac{2\overline{u_i'^2}}{\lambda^2} \tag{4.5}$$

Rearranging, and noting that the Nyquist sampling theorem dictates that to avoid aliasing the sampling frequency (f) should be twice the frequency of the smallest fluctuations (i.e., those with the shortest timescales),

$$f = \frac{2}{\lambda} = \sqrt{2\overline{\left(\frac{du_i'}{dt}\right)^2} \Big/ \overline{u_i'^2}} \tag{4.6}$$

In addition, Tennekes and Lumley (1972) use (4.3) to define the integral scale $\mathfrak{I}$.

$$\mathfrak{I} \equiv \int_0^\infty \alpha(\Delta t)d\Delta t \tag{4.7}$$

Noting that it is assumed that $\mathfrak{I}$ is always finite, $\mathfrak{I}$ is considered a measure of the time period over which u_i' is correlated with itself. Tennekes & Lumley (1972) then show that the mean-square error in an estimate of the time-averaged velocity may be approximated by:

$$\overline{\left(\overline{U_i} - \overline{u_i}\right)^2} \cong 2\overline{u_i'^2}\,\Im/T \tag{4.8}$$

where the overbar uppercase U_i denotes the estimate of the time-averaged velocity over the finite sampling period T. Thus, if the integral scale $\Im$ is finite, the error in the time average is inversely proportional to the sampling period and the estimated time average converges to its true value as sampling period increases. Tennekes & Lumley (1972) conclude that averages converge and integral scales exist if $T = 2\Im$. More recent work by Pope (2000) suggests that $T = 8\Im$.

The preceding paragraphs discuss the time scales inherent in turbulent flows and identify rigorous methods for selecting the optimum sampling frequency and sampling period. In some cases, however, a suitable sampling period can be defined by the physical constraints of the flow domain. For example, in pipe flow the diameter of the pipe is of the order of the largest eddies in the flow and the ratio of the pipe diameter to mean velocity along the pipe is a good estimate of the time period required to describe the flow. To apply this approach to natural flows, assume that a turbulent boundary layer originates from a source of roughness. This may be, for example, a gravel particle, a sessile animal, a plant or stand of plants. The mean period of time for fluid to travel some characteristic length, L, away from this source is $L/\overline{u_i}$. At L, the boundary layer has a thickness l; the time rate of change of boundary layer thickness, dl/dt, is approximated by $\overline{u_i'^2}$ the root mean square of the velocity fluctuations associated with turbulent eddies. As the periods of elapsed time for fluid to travel the distance L and for the boundary layer to expand to a thickness l are the same, it is possible to express:

$$\Im \sim l/\overline{u_i'^2} \sim L/\overline{u_i} \tag{4.9}$$

Thus, if it is observed that eddies being shed from a source travel 5 m from that source before they break the water surface and the fluid is flowing at ~0.5 m s^{-1}, following Pope (2000) the sampling period should be ~80 s. Further, Tennekes & Lumley (1972) use results from Taylor (1935) and Kolmogorov (1941) to estimate the timescale of the smallest eddies as:

$$\lambda = \frac{l}{\overline{u_i'^2}}\left(\frac{\nu}{\overline{u_i'^2}l}\right)^{1/2} = \Im\,\mathrm{Re}^{-1/2} \tag{4.10}$$

where Re is the Reynolds number calculated from the thickness of the boundary layer, the root mean square of velocity fluctuations and the kinematic viscosity, ν, of the fluid. Hence, once again noting that the Nyquist sampling theorem dictates that to avoid aliasing, the sampling frequency should be twice the frequency of the smallest fluctuations (i.e., those with the shortest timescales),

$$f = \frac{2}{\lambda} = 2\frac{\overline{u_i'^2}}{l}\left(\frac{\overline{u_i'^2}l}{\nu}\right)^{1/2} = 2\frac{\mathrm{Re}^{1/2}}{\Im} \tag{4.11}$$

4.1.2   Measurement uncertainties in turbulent flows

Uncertainties in velocity measurements tend to be of two kinds: random and systematic (Taylor, 1997). Random errors tend to be inherent in any set of measurements and, in the absence of systematic errors, cause successive measurements of a variable to vary about its true value (Squires, 1968). The distribution of measurements about this value is commonly assumed to be Gaussian. Conversely, systematic errors are regular or repeated throughout a set of measurements (Squires, 1968), such that measurements are consistently displaced from the true value. Soulsby (1980) described a number of potential sources of error in the measurement of turbulent flows, some of which have been mentioned above. These are:

1   loss of low-frequency fluctuations in the downstream velocity. Selection of too short a record prevents sampling of the largest turbulent motions, whose time scales are longer than the sampling period;
2   large random error in the covariance of the downstream and vertical velocities. The random error associated with estimation of the mean or variance is inversely proportional to the sampling period (equation 4.8). Thus, to minimise random error, record length must be maximised;
3   error introduced during the calculation of mean velocities by flow unsteadiness during the recording period. In areas in which the mean flow varies, such as coastal regions where the flow is driven primarily by the tide, the record length should be sufficiently short with respect to the predominant tidal half-period so that, at the very least, the flow can be considered as quasi-stationary. Alternatively, if the timescale of the flow unsteadiness is similar to the integral timescale, the effects of unsteadiness may be reduced by applying, for example, a low pass filter or by phase-averaging the velocity records; and
4   loss of high-frequency fluctuations in the vertical velocity, caused by either the application of an instrument with too low a sampling frequency or too large a sampling volume. The selection of an instrument with a low sampling frequency relative to the frequency distribution of turbulent motions prevents sampling of the smallest turbulent motions. Alternatively, the velocity measured by the instrument is a spatial average over the sampling volume, weighted by a function specific to the sensor. For two simplified weighting functions, one a top-hat function and the other a modified Gaussian, Soulsby (1980) showed that the loss of high frequency fluctuations was sensitive to a scale describing the size of the sampling volume but insensitive to the form of the weighting function.

It is evident that a balance must be struck between these factors as some are diametrically opposed. For the specific tidal environment (Start Bay, Devon, UK) studied by Soulsby (1980), it was found that to limit the loss of low-frequency fluctuations to 5%, the record length needed to be 840 s, to characterise burst and sweep phenomena required a record length of 1980 s and to limit random error to 10%, the record length needed to be 5760 s (96 minutes). Conversely, owing to variability in tidal cycles, the longest period of time that exhibited quasi-stationary behaviour was only 480 s. During a study of bottom currents in 19–21 m deep areas of the New York Bight, Lesht (1980) argued that the record length could be as short as 20 times the characteristic time scale of the flow, which he estimated to be six seconds, yielding a recommended minimum record length of 120 s. Conversely, from a survey of 50 fluvial studies, Buffin-Bélanger & Roy (2005) reported that 35% used a record length of 60 s and 65% used a record length shorter than 120 s. Selection of record length by fluvial scientists was not related to sensor characteristics, sampling frequency, flow depth nor turbulence intensity but rather, to the objectives of the study, the type of data analyses to be conducted on the velocity records and also to time constraints associated with the survey of several sampling points during a given period of time (Buffin-Bélanger & Roy, 2005). These authors showed that analyses in the time domain, such as spectral and cross-correlation functions, usually relied on a few long records, whereas studies focusing on the spatial description of the flow structure compromised record length for the benefit of larger spatial distribution. Sukhodolov & Rhoads (2001) used the convergence of the cumulative velocity variance associated with different time-averaging windows to conclude that record length should be 60 s in the ambient flow approaching a confluence, but extended to 90–100 s in a region of higher turbulence intensity. Buffin-Bélanger & Roy (2005) used bootstrapping techniques, which allowed distributions of 10 commonly-used turbulence statistics to be computed from 19 time series of different lengths, to estimate standard errors. They concluded that the optimum number of samples was dependent upon the velocity component considered and also the order of the turbulence statistic; for the mean, turbulence intensity and skewness of the streamwise velocity component, the optimal numbers of samples were 1250, 1280 and 1640, respectively. For the vertical component, the optimal numbers of samples were 1090, 1250 and 1550, respectively. Crucially, Buffin-Bélanger & Roy (2005) noted that errors were a non-varying function of the number of velocity samples, not the period of time, comprising a flow record (note that the record lengths computed by Soulsby (1980) equate to 1680, 3960 and 11520 samples, respectively). However, for the two instruments employed in their study, optimal durations ranged between 60 and 90 s for most statistics (Buffin-Bélanger & Roy, 2005).

Table 4.1 lists ecohydraulic studies which have used either Acoustic Doppler Velocimetry (ADV), Hot-Wire or Hot-Film Anemometry (both denoted HWA) to sample velocities. It includes the purpose of the study, the quoted operating frequency of the instrument and the duration and number of samples. The median number of velocity samples measured by researchers varies substantially, from 20–15000, with a median of 3000. This median may represent a rough rule of thumb for the number of samples required to quantify turbulent flows interacting with ecology. The table clearly suggests that some previous studies have sampled velocities for too short a period of time. For example, the error introduced into the measurements of just 1–5 s duration by Friedrichs & Graf (2009) may be 11–48 times the error of a 60 s sample, or 20–88 times the error of a 120 s sample (Buffin-Bélanger & Roy, 2005; their Fig. 8d). Thus, measurements taken for short durations may be unreliable.

4.1.3 Problems of measuring flows within and around live biota

Turbulent boundary layer flows are highly irregular in space and time, and are thus extremely complex. Measuring turbulent flows presents a number of problems that must be overcome, or, at least, mitigated:

Table 4.1 Sampling parameters from 30 ecohydraulic studies conducted using Acoustic Doppler Velocimetry (ADV) or Hot-Wire or Hot-Film Anemometry (both denoted HWA) to measure velocity time series.

Reference	Purpose	Instrument	Sampling frequency (Hz)	Sampling duration (s)	Number of samples
Ghisalberti & Nepf (2002)	Velocity profiles in a surrogate flexible eelgrass meadow	ADV	25	600	15000
Nepf (1999)	Velocity profiles in an array of wooden cylinders	ADV	25	360	9000
Nepf & Vivoni (2000)	Velocity profiles in a surrogate flexible eelgrass meadow	ADV	25	360	9000
Hendriks et al. (2006)	Structure of boundary layer and turbulence above smooth sediment bed and ridge of pacific oysters	ADV	25	330	8250
Van Duren et al. (2006)	Structure of boundary layer and turbulence above a bed of live mussels (Mytilus edulis)	ADV	25	330	8250
Bouma et al. (2007)	Velocity profiles and turbulence statistics through an array of bamboo sticks	ADV	25	330	8250
Wallace et al. (1998)	Mean velocities and turbulence statistics within an artificial seagrass canopy	ADV	25	300	7500
Nikora et al. (2002b)	Mean velocities and turbulence statistics within periphyton communities dominated by the diatom Synedra and the green filamentous Spirogyra	ADV	100	60	6000
Larned et al. (2004)	Mean velocities and turbulence statistics within assemblages of periphyton	ADV	50	120	6000
Zong & Nepf (2010)	Longitudinal velocity transects through and at the edge of an array of cylinders	ADV	25	240	6000
Wilson & Horritt (2002); Wilson et al. (2003, 2005)	Velocities within and above real and two types of artificial grass canopies; Velocity and turbulence statistics within an artificial Laminaria hyperborea stand	ADV	25	180	4500
Finelli et al. (2002)	Effects of velocity and particle concentration on the posture, flick rate, and ingestion rate of Simulium vittatum larvae	HWA	20	204.8	4096
Salant (2011)	Velocity and turbulence profiles in and above periphyton assemblages dominated by diatoms (Achnanthes minutissima) and filamentous green algae (Rhizoclonium riparium).	ADV	50	60–90	3000–4500

Reference	Description	Instrument			
Leonard & Luther (1995)	Velocities through *Juncus roemerianus, Spartina alterniflora* and *Distichlis spicata*	HWA	5	600–1920	3000–9600
Neumeier (2007)	Velocity/turbulence profiles around three canopies of *Spartina anglica*	ADV	25	120	3000
Seraphin & Guyenne (2008)	Velocity and turbulence profiles in and above stands of rigid surrogates	HWA	90	30	2700
Folkard & Gascoigne (2009)	Velocity and turbulence profiles above patches of dead mussels (*Mytilus edulis*)	ADV	25	90	2250
Järvelä (2005)	Velocity profiles and associated turbulence statistics in a stand of wheat	ADV	25	60–120	1500–3000
McBride et al. (2007)	Velocity fields within synthetic grass carpeting and randomly-distributed rigid wooden dowels	ADV	25	60–120	1500–3000
Thomson et al. (2004)	Flow patterns downstream from individual *Simulium vittatum* larvae	HWA	128	10	1280
Sand-Jensen & Mebus (1996)	Flow patterns within patches of four dominant macrophyte species from Danish lowland streams	HWA	20	50	1000
Lassen et al. (2006)	Concentration, velocity and turbulence profiles above beds of *Mytilus edulis*	ADV	20	30	600
Biggs et al. (1998)	Velocity profiles around periphyton-covered cobbles	HWA	10	50	500
Jones et al. (2012)	Velocity profiles (short sampling duration) and near bed shear stress (long sampling duration) above *Austrovenus stutchburyi* (an infaunal clam)	ADV	25	20–163	500–4075
Friedrichs et al. (2000)	Vertical and horizontal velocity profiles in artificial polychaete tube lawns	ADV	15	30	450
Cornelisen & Thomas (2006)	Velocities and nutrient uptake in meadows of *Thalassia testudinum*	ADV	5	60	300
Morris et al. (2008)	Velocity and turbulence patterns in stands of two seagrass species (*Zostera noltii* and *Cymodocea nodosa*)	ADV	25	5	125
Friedrichs & Graf (2009)	Velocities and turbulent structures around replica macrozoobenthic species	ADV	20	1–5	20–100

- Turbulent flows are four-dimensional in nature, with variability in all three spatial dimensions as well as time. Thus, the first challenge is to (ideally simultaneously) sample velocities at numerous locations that are distributed in the streamwise, cross-stream and vertical directions;
- Turbulent flows, especially those concerned with a boundary layer, are generally not isotropic. That is, turbulent motions tend to be larger and stronger in one direction than the others. Thus, the second challenge is to simultaneously sample more than one velocity component. This will allow reasonable estimates of turbulent stresses to be computed;
- Turbulent flows exhibit a range of spatial scales, from the smallest eddies defined by the Kolmogorov (1941) microscale ($[\nu^3/\varepsilon]^{1/4}$, where ε is the dissipation rate per unit mass, which can be approximated by Taylor's (1935) relation $\overline{u_i'^2}^3/l$) to the largest eddies defined by the thickness of the boundary layer. The third challenge is therefore to adequately sample fluctuations across this range of spatial scales to enable the spatial visualisation of flow structures;
- Turbulent flows exhibit a range of temporal scales (Section 4.1.1) and thus the fourth challenge is to sample at a rate that is fast enough to detect the smallest scales, but to do so for a sufficient period to ensure accurate and precise estimates of time-averages (i.e. to ensure convergence). This will enable the temporal visualisation of flow structures;
- Turbulent flows interact with structures within the fluid, potentially altering local flow patterns. Thus, the fifth challenge is to employ ideally non-, or at worst minimally-, intrusive devices.

Measuring turbulent flows within and around live biota adds to these challenges. First, measurements should not interfere with the normal behaviour of the organism being studied, such that flow measurements are unrealistic. Second, some of the most interesting ecohydraulic problems involve biota that are capable of moving, either in response to fluid flows or to other stimuli. Thus, measurement techniques should be capable of sampling close to and around moving biota, for example plant canopies. Third, many organisms exist in close proximity to each other and therefore present a region which is porous to the flow field, with some regions of space occupied by relatively solid objects and others occupied by the fluid. Thus, measurement techniques should ideally be capable of sampling close to and within these complex structures, for example plant canopies, polychaete tube lawns, or colonies of filter feeders such as oysters, barnacles and mussels.

4.1.4 Technique overview

An overview of the techniques described and reviewed in this section is provided in Table 4.2 and expanded within the following sections. The focus is on techniques that can quantify high frequency turbulent velocity fluctuations. Thus, no consideration is made of studies that have employed electromagnetic (e.g. Fonseca & Fisher, 1986; Pilditch & Grant, 1999; Freeman et al., 2000; Schanz & Asmus, 2003; Sand-Jensen, 2003, 2008; Peterson et al., 2004; James et al., 2008) or portable propeller (e.g. Fathi-Maghadam & Kouwen, 1997; Freeman et al., 2000; Shi & Hughes, 2002) flow meters that are incapable of sampling velocities at rates faster than 2 times per second (2 Hz)

Table 4.2 Summary of flow measurement techniques reviewed in this chapter.

Technique	Key references	Applications to ecohydraulics	Strengths	Weaknesses
ACOUSTIC TECHNIQUES				
Acoustic Doppler Velocimetry (ADV)	Lhermitte (1983); Kraus et al. (1994); Lohrmann et al. (1995); Voulgaris & Trowbridge (1998); McLelland & Nicholas (2000)	Dunn et al. (1996); Nepf (1999); Nepf & Vivoni (2000); Friedrichs et al. (2000); Lopez & García (2001); Ghisalberti & Nepf (2002); Nikora et al. (2002b); Wilson & Horritt (2002); Wilson et al. (2003, 2005); White & Nepf (2003); Larned et al. (2004); Järvelä (2005); Cornelisen & Thomas (2006); Hendriks et al. (2006); Lancaster et al. (2006); Rice et al. (2008); Bouma et al. (2007, 2009a, 2009b); McBride et al. (2007); Neumeier (2007); Morris et al. (2008); Friedrichs & Graf (2009); Zong & Nepf (2010); Salant (2011)	"Plug and play". Setup is straightforward; Relatively inexpensive (~€7500); Three velocity components Sampling rates 0.1–200 Hz; Relatively small sampling volume; Sampling volume is displaced from the transducer to reduce interference; Claimed accuracy of the probes is ±1.0%	Cannot be used in extremely shallow water; Probe acts as an obstacle; Single point velocities; Probes must be carefully orientated; Errors can be caused by: velocity gradients and shearing within the sampling volume; proximity to a boundary; aerated flows and highly turbulent zones; Data must be post-processed before interpretation; Reynolds stress are overestimated at velocities <0.1 m s^{-1}

(Continued)

Table 4.2 (Continued).

Technique	Key references	Applications to ecohydraulics	Strengths	Weaknesses
Acoustic Doppler Velocity Profiling (ADVP)	Takeda (1991); Lhermitte & Lemmin (1994); Shen & Lemmin (1997); Lemmin & Rolland (1997); Rolland & Lemmin (1997); Hurther & Lemmin (1998, 2001); Blanckaert & Lemmin (2006)	Graf & Yulistiyanto (1998); Graf & Istiarto (2002); Garcia et al. (2011)	Multiple measurements of velocity are made quasi-simultaneously along a transect; Sampling rates 0.1–200 Hz; Level of intrusion depends on choice of ADVP or UDVP and if the latter on the chosen velocity component; Ultrasound is penetrative so is theoretically capable of measuring within plant canopies and soft-tissued fauna; Tolerant of large volumes of suspended sediment in the water column	ADVP is not presently commercially available although Nortek have developed profiler firmware for the Vectrino; UDVP can only measure one component velocities; Probes must be carefully orientated; Errors can be caused by: velocity gradients and shearing within the sampling volume; proximity to a reflective boundary; aerated flows and highly turbulent zones; Ultrasonics needs more turbid water to operate optimally; Optimal conditions for UDVP measurements may far exceed acceptable levels for biota; Data must be post-processed before interpretation;

OPTICAL TECHNIQUES

Laser Doppler Anemometry/ Velocimetry (LDA/LDV)	Nezu & Rodi (1986); Durst et al. (1981); Adrian (1996)	Non-intrusive; Three component velocities; Very high frequency response (>100 Hz); Very small sampling volume (<1 mm^3); Extremely accurate	Gambi et al. (1990); Nepf (1999); Nepf & Vivoni (2000); Nezu & Onitsuka (2001); Ghisalberti & Nepf (2002); White & Nepf (2007, 2008); Choi et al. (2009); Graba et al. (2010); Liu et al. (2010)	Very expensive; Laser alignment is non-trivial and time-consuming; Single point velocities; Line-of-sight required between the laser, the measurement volume and the photodetector; Reduced performance in turbid water; Cannot measure shallow flows; Potentially damaging to both flora and fauna
Particle Image Velocimetry (PIV)	Keane & Adrian (1992); Westerweel (1994, 1997); Adrian (1996); Raffel et al. (1998)	Non-intrusive; Two-component velocities (basic), three-component velocities with stereoscopic PIV; Very high frequency response (potentially >1000 Hz); Very high spatial resolution, limited only by the camera lens and pixel size/resolution of the camera;	Stamhuis & Videler, (1998a, 1998b, 1998c); Müller et al. (2000); Nezu & Onitsuka (2001); Stamhuis et al. (2002); van Duren et al. (2003); Schindler et al. (2004); Sakakibara et al. (2004); Nagayama & Tanaka (2006); Hultmark et al. (2007); Moulin et al. (2007); Nagayama et al. (2008); Tinoco & Cowen (2009)	Very expensive; Non-trivial setup; Enormously hard disk space- and memory-intensive; Line-of-sight required between the light source, the camera, and the measurement volume; Fluid must be relatively clear;

(Continued)

Table 4.2 (Continued).

Technique	Key references	Applications to ecohydraulics	Strengths	Weaknesses
			Can be used in extremely shallow water	Errors can be caused by: inaccurate synchronisation of laser/light source pulses and image capture; Density of seeding is significantly different to that of the fluid; Variation in grain size and light reflective/absorptive qualities of seeding material; Potentially damaging to both flora and fauna, although possible to use lower energy light sources
THERMAL TECHNIQUES				
Hot-Wire Anemometry (HWA)	Hanratty & Campbell (1983); Gust (1988); Fernholz et al. (1996)	Gambi et al. (1990); Leonard & Luther (1995); Sand-Jensen & Mebus (1996); Biggs et al. (1998); Finelli et al. (2002); Thomson et al. (2004); Seraphin & Guyenne (2008)	Very high frequency response (up to 400 kHz); Small sampling volume (<1 mm³)	Expensive; Usually one or two components, although some three component devices are available; Single point velocities; Probe acts as an obstacle; Insensitive to slow velocities and cannot measure negative velocities; Unsuitable for moving objects; Very sensitive to fouling: require frequent recalibration and/or cleaning; Hot probe may cause trauma to the plant or animal

or over an area that is small enough for computed measures of turbulent intensity to be meaningful. Furthermore, no consideration is made of studies that have timed the downstream advection of tracer dye (e.g. Yund & Meidel, 2003) as these cannot quantify turbulence and also have questionable accuracy and precision.

4.2 ACOUSTIC TECHNIQUES: THE DOPPLER SHIFT

Currently-available acoustic flow measurement techniques capable of sampling at rates sufficiently fast to quantify turbulence rely upon the Doppler shift. If a source of sound is moving relative to a receiver, the frequency of sound at the receiver is shifted from the transmitted frequency by the amount

$$f_D = -2f_0 \left(\frac{u}{c} \right) \qquad (4.12)$$

where f_D is the change in received frequency, or Doppler-shift frequency; f_0 is the frequency of the transmitted sound; u is the velocity of the source relative to the receiver; and c is the speed of sound. After rearrangement, it is thus possible to obtain an equation for u:

$$u = \frac{-cf_D}{2f_0} \qquad (4.13)$$

4.3 ACOUSTIC TECHNIQUES: ACOUSTIC DOPPLER VELOCIMETRY (ADV)

4.3.1 Introduction to ADV

The Acoustic Doppler Velocimeter (ADV) is a single point, three-component Doppler current meter. An acoustic pulse is emitted from the centre of a claw-like transmitter-receiver array, and reflected echoes are detected by 3–4 receivers. The geometry of the most common receivers means that the cylindrical sampling volume is displaced some 50 mm from the array. The sampling volume varies in size depending upon the unit; older ADVs tend to have sampling volumes that are 9 mm in height and 6 mm in diameter, with a volume of ~254–346 mm³, while more recent "laboratory" ADVs have sampling volumes that are 3 mm in height and 6 mm in diameter, with a volume of ~85 mm³.

Typically, a single velocity measurement is calculated as an average of 150–250 returning dual-pulse pairs per second (where the rate varies with velocity range setting). Each returning dual-pulse pair, or ping, incorporates Doppler noise, and therefore pings are averaged to reduce the noise level in each velocity value. For the most commonly used instruments, the number of pings averaged to produce a single velocity estimate is equal to the sampling rate, which may be set at 0.1–200 Hz. In addition to velocities, the ADV software also provides the Signal Amplitude (S), the Signal-to-Noise Ratio (SNR) and the correlation between successive radial velocities (R^2) for each receiver. This additional information can be used to assess the quality of the data. Because the acoustic signal is contaminated with noise that is intrinsic to the propagation of acoustic waves and the Doppler detection system, raw ADV velocity data should not be used to compute turbulence statistics without adequate post-processing (e.g. McLelland & Nicholas, 2000; Goring & Nikora, 2002; Wahl, 2003; García et al., 2005; Chanson et al., 2007; Doroudian et al., 2007).

In steady flows, the first stage of ADV signal processing is the removal of all data samples with communication errors (average correlation below 60% or Signal-to-Noise Ratio (SNR) below 5 to 15 dB, McLelland & Nicholas, 2000). As described by McLelland & Nicholas (2000), the covariance method is used to estimate the correlation between dual pulse-pair velocity estimates. However, the spectral width (or spectral variance) of the covariance function is contaminated by noise due to the loss of signal coherence during its propagation through the fluid (Lhermitte & Serafin, 1984). This white noise cannot be removed by improving the acoustic qualities of the system, but different pulse repetition rates permit the separation of variance caused by the signal from that contributed by noise (Lhermitte & Serafin, 1984). Assuming a Gaussian distribution for the phase power spectrum, the correlation coefficient, R^2, can be related to the dimensionless spectral width, φ_r:

$$R^2 = e^{-2\pi^2 \varphi_r^2} \tag{4.14}$$

where e is the base of natural logarithms. The dimensionless spectral width is the product of the received signal width and the sample time interval. Hence, for a given correlation coefficient the variance of the received signal will increase for larger instrument velocity ranges because the sample time interval decreases as the velocity range increases. If the signal spectrum conforms to a Gaussian distribution, R^2 should exceed ~0.6 ($\varphi_r = 0.16$) to prevent aliasing of the mean velocity estimated from the received signal spectrum (Lhermitte & Serafin, 1984). At this R^2, the difference between the measured spectral variance and the variance for different spectral shapes (bimodal, square or exponential distributions) is less than 8% (Lhermitte & Serafin, 1984).

Velocity samples with Signal-to-Noise Ratio (SNR) below 5–15 dB should be removed (McLelland & Nicholas, 2000). SNR is calculated using the signal amplitude, S, and background noise level, N, and can be expressed in decibels as:

$$SNR = 0.43 \, (S - N) \tag{4.15}$$

Miller & Rochwarger (1972) and Zrnic (1977) established that the standard deviation of mean velocity estimated from independent pulse-pairs is controlled by both SNR and φ_r. For a given instrument velocity range, the dimensionless spectral width of the signal noise, φ_n^2, is:

$$\varphi_n^2 = \varphi_{rn}^2 - \varphi_r^2 = \frac{\ln[1 + (1/SNR)]}{2\pi^2} \tag{4.16}$$

where φ_{rn}^2 is the combined spectral width of the signal and noise spectra. SNR should be maximized during measurements; Nortek (1997) suggest that SNR >5 dB is required for collecting mean flow data and that SNR >15 dB is necessary for collecting instantaneous flow data.

The remaining data should then be "despiked" using the phase-space thresholding technique (e.g. Goring & Nikora, 2002; Wahl, 2003), or another unbiased error-estimator. The phase-space thresholding technique identifies erroneous data through applying the observation that good ADV data are tightly clumped within an ellipsoid in phase space (three-dimensional plots of velocity, u_i, and approximations of its first and second derivatives, du_i/dt and d^2u_i/dt^2, respectively; Goring & Nikora, 2002).

4.3.2   The strengths and weaknesses of ADV

Strengths:

- ADVs are essentially "plug and play". Setup is straightforward and does not require expert input;
- ADVs are relatively inexpensive (~€7500);
- Three to four sensing heads measure velocities in three orthogonal directions. This allows the calculation of three-component turbulence statistics;
- Sampling rate ranges of 0.1–200 Hz are now common in modern probes;
- The sampling volume is relatively small: ADVs have sampling volumes that are 6 mm in diameter and 3–9 mm in height, with volumes ranging from ~85–346 mm³;
- The sampling volume is displaced a distance of ~50 mm from the transducer array, reducing the physical influence of the probe upon the flow. Interference can be reduced further through the use of sideways-looking probes rather than downward-looking probes;
- Claimed accuracy of the probes is ±1.0% (Lohrmann *et al.*, 1995; Voulgaris & Trowbridge, 1998);
- Acoustic backscatter data may also be used to provide information regarding the quantity and type of particulate matter in the flow.

Weaknesses:

- Finite blanking distance between the probe and the sampling volume (~50 mm) and finite required submergence depth (10–80 mm) prevent deployment of downward-looking probes in extremely shallow water. However, sideways-looking probes overcome this issue;
- Although the sampling volume is not directly beneath the probe, the ADV is still semi-intrusive because it is only ~50 mm from the area of interest. It is therefore an obstacle to the flow and disturbs flow characteristics at the measurement point (Strom & Papanicolaou, 2007), to an as yet un-quantified degree;
- Commercially-available ADVs provide three component velocities at a single point and cannot provide simultaneous measurements of velocities throughout the water column. Thus, in order to obtain time-averaged velocity profiles, either multiple probes need to be deployed simultaneously or multiple point measurements must be combined, increasing the total period of time over which measurements are made. However, recent developments by Nortek do permit the

simultaneous measurement of velocities in up to 35 sampling volumes, each of 1 mm in height;

- Probes must be carefully orientated so that the three component velocities are located in the desired coordinate system and can be converted to three-dimensional velocities;
- Errors can be caused by velocity gradients and shearing within the sampling volume (Strom & Papanicolaou, 2007) and irresolvable eddies that are smaller than the sampling volume (e.g. those shed by biofilms or by vegetation with thin stems; Chapters 5 and 6; Barkdoll, 2002);
- Errors can be caused by proximity to a boundary (e.g. the bed, (Finelli et al., 1999) or biota). Precht et al. (2006) and Chanson et al. (2007) note that both R^2 and SNR decrease significantly with increasing proximity to solid boundaries. Using very flexible sea grass surrogates of 2 mm width at densities as low as 500 shoots m^{-2}, Paul (personal communication, October 2011) found that data quality was too poor for quantitative analysis. This results in the underestimation of the streamwise velocity component when the boundary is less than 30 to 45 mm from the sampling volume (Chanson et al., 2007);
- Errors can be caused by measurements in aerated flows and highly turbulent zones (Cea et al., 2007; Strom & Papanicolaou, 2007);
- Data must be post-processed before interpretation. However, freely available software packages (e.g. ExploreV, WinADV) are available to do this;
- Lohrmann et al. (1995) report that at low velocities (<100 mm s^{-1}), ADV measurements of Reynolds stress are overestimated because of variation in the sensitivity of the three ADV receivers. However, this should not be a problem at higher velocities.

4.3.3   The use of ADVs in ecological studies

Since their introduction in the early 1990s, ADVs have been used extensively to quantify the flow fields in the vicinity of plants and animals and their surrogates. For example, Nikora et al. (2002b), Larned et al. (2004) and Salant (2011) used ADVs to measure mean and/or turbulent velocity fields within and around live biofilms mats. Wilson & Horritt (2002), Järvelä (2005), Cornelisen & Thomas (2006), Neumeier (2007), Bouma et al. (2009a, 2009b) and Morris et al. (2008) used them to measure velocity fields around stands of live vegetation. Dunn et al. (1996), Nepf (1999), Nepf & Vivoni (2000), Ghisalberti & Nepf (2002), Wilson et al. (2003, 2005), White & Nepf (2003), Bouma et al. (2007, 2009a, 2009b), McBride et al. (2007), and Zong & Nepf (2010) studied velocity fields around artificial vegetation. Hendriks et al. (2006) used ADVs to characterise the structure of the boundary layer and turbulence above a smooth sediment bed and above a ridge of live pacific oysters (*Crassostrea gigas*) as well as the impact of changing roughness on the behaviour of larvae. Lassen et al. (2006) used ADV measurements to compare turbulent flow properties during different filtration activities of blue mussels (*Mytilus edulis*). Folkard & Gascoigne (2009) measured the impact of varying incident flow velocities and heterogeneity in spatial distributions of *Mytilus edulis* on the form of velocity profiles and the strength of turbulent structures. Friedrichs and co-workers (Friedrichs et al., 2000; Friedrichs & Graf, 2009) measured vertical and horizontal velocity profiles around replica macrozoobenthic features: polychaete worm tubes, a snail shell, a mussel, a sand mound, a pit, and a cross-stream track furrow (see Table 4.1 for additional examples).

4.4 ACOUSTIC TECHNIQUES: ACOUSTIC DOPPLER VELOCITY PROFILING (ADVP)

4.4.1 Introduction to ADVP

The Acoustic Doppler Velocity Profiler (ADVP) measures quasi-instantaneous three component velocities through an entire water column segmented into finite measurement volumes 30–50 mm in height. It consists of a central emitter surrounded by four receivers and placed in a separate water-filled housing that is in contact with the water surface, minimising intrusion. The emitter is focalised and has a constant beam diameter of ~70 mm (Hurther & Lemmin, 1998). Alternatively, the Ultrasonic Doppler Velocity Profiler (UDVP) provides quasi-instantaneous one component velocities through an entire water column; UDVP transducers are partially focalised such that the beam diameter expands at an angle of ~1.3–3.3° from the transducer. Dependent upon the velocity component, UDVP may be intrusive or non-intrusive; cross-stream and vertical velocities may be measured through flume walls using ultrasonic coupling gel, but streamwise velocities usually require immersion within the fluid. Both ADVP and UDVP rely upon the transmission of an ultrasonic burst of sound (for the ADVP, f_0 is constant at 1 MHz; 0.5–8 MHz transducers are presently available for the UDVP) from a piezoelectric transducer along a measurement profile. This burst propagates through the fluid and is reflected from the surface of microparticles suspended in the liquid before being received. The spatial and velocity information of the suspended particles assumed to be travelling with the velocity of the fluid flow, and hence the velocity profile, is contained in the reflected waves (echoes). The distance between a suspended particle and the transducer (x) is calculated from the time delay (t) between the start of the burst and the reception signal,

$$x = \frac{ct}{2} \tag{4.17}$$

where c is the velocity of ultrasound in the fluid being investigated (assuming the density of the fluid in the measurement area remains constant).

The return signal (echo) is gated along the measurement axis at known return times, allowing the measurement of velocity at (usually) 128 separate, but evenly spaced, positions along the ultrasound emission axis. The maximum distance to which the instruments will detect, L_{max} and the maximum detectable velocity, U_{max} are determined by:

$$L_{max} = \frac{c}{2f_{pr}} \tag{4.18}$$

and

$$U_{max} = \frac{cf_{pr}}{4f_0} \tag{4.19}$$

respectively, where f_{pr} is the pulse repetition frequency (the number of acoustic wave pulses generated by the emitter per second, or sampling frequency, dictated by the Nyquist sampling theorem such that $f_{pr} > 2f_D$).

Different aspects of both ADVP (Lhermitte & Lemmin, 1994; Shen & Lemmin, 1997; Lemmin & Rolland, 1997; Rolland & Lemmin, 1997; Hurther & Lemmin, 1998, 2001) and UDVP development (Takeda, 1991) have been reported in the literature. The ADVP employs the same pulse-pair algorithm (Lhermitte & Serafin, 1984) to calculate velocities as the ADV. As with all acoustic techniques, the acoustic signal is contaminated by noise and thus data should be carefully post-processed (e.g. Buckee *et al.*, 2001 for UDVP and Hurther & Lemmin, 2001; Franca & Lemmin, 2006; Blanckaert & Lemmin, 2006 for ADVP).

4.4.2   Strengths and weaknesses of ADVP

Strengths:

- Multiple measurements of velocity are made quasi-simultaneously, reducing experimental time and allowing spatial velocity gradients to be computed along the measurement axis;
- Sampling rate ranges of 0.1–200 Hz are possible;
- The ADVP is minimally intrusive, with pressure exerted on the water surface by the weight of the water-filled housing;
- The UDVP is either non-intrusive or intrusive, depending upon the chosen velocity component;
- Ultrasound is penetrative within liquids and lightweight tissues. Thus, both ADVP and UDVP are theoretically capable of measuring within plant canopies and soft-tissued fauna;
- Both ADVP and UDVP are more tolerant of large volumes of suspended sediment in the water column (Buckee *et al.*, 2001) than ADV or optical techniques. Thus, they may be deployed in more turbid environments;
- The ADVP has a lower signal-to-noise ratio than a traditional three-receiver ADV because of the use of four receivers. It has a similar SNR to modern four-receiver ADV units;
- Acoustic backscatter data may also be used to provide information regarding the quantity and type of particulate matter in the flow.

Weaknesses:

- ADVP was not commercially available at the time of writing although, as noted previously, Nortek have developed firmware to permit the simultaneous

measurement of velocities in up to 35 sampling volumes, each of 1 mm in height. However, this creates a rather short measurement column;
- UDVP can only measure one component velocities;
- Probes must be carefully orientated so that the three component velocities measured by the ADVP and the one component velocities measured by the UDVP are located in the desired coordinate system and can be converted to dimensional velocities;
- Errors can be caused by velocity gradients and shearing within the sampling volumes (Strom & Papanicolaou, 2007) and irresolvable eddies that are smaller than the sampling volumes (e.g. those shed by biofilms or by vegetation; Chapters 5 and 6; Barkdoll, 2002);
- Errors can be caused by proximity to a reflective boundary (e.g. metals, PVC; Graf & Yulistiyanto, 1998);
- Errors can be caused by bubbles in aerated flows and highly turbulent zones (Cea et al., 2007, Strom & Papanicolaou, 2007);
- Ultrasonic devices are commonly expected to operate optimally in rather turbid fluids. Thus, the optimal conditions for UDVP measurements may far exceed acceptable levels for biota;
- Data must be post-processed before interpretation.

4.4.3   Use of ADVP in ecological studies

At the time of writing, data collected using ADVP or UDVP to study ecohydraulics had not been reported in the literature, although preliminary results using an ADVP to study invertebrate drift were reported in 2011 (Garcia et al., 2011). The ADVP has been successfully applied in various studies on straight, uniform and non-uniform clear water flows over rigid as well as mobile bottom configurations (Song et al., 1994; Song & Graf, 1996), straight, uniform suspension flow (Cellino & Graf, 1999), highly three-dimensional curved flow (Blanckaert, 2009, 2010; Blanckaert & Graf, 2001, 2004; Blanckaert & de Vriend 2004, 2005a, 2005b), and turbulent energy production and dissipation in fully rough gravel-bed open-channel flows (Mignot et al., 2009). Graf & Yulistiyanto (1998) used an ADVP sampling at 12 Hz to study the flow and turbulent structures around a single 220 mm-diameter cylinder placed on an immovable bed in a 2 m wide flume. They were able to visualise a horseshoe-vortex system that stretched and thus decayed in strength as it wrapped around the base of the cylinder. Downstream, flow separation was detected, with significant flow reversal and an increase in the turbulence intensities. Graf & Istiarto (2002) followed this work using an ADVP, sampling at 20–28 Hz, to study the impact of the horseshoe-vortex system on an erodible bed with a 2.1 mm mean grain size. The deformed bed around the scour hole prompted the formation of another, weaker, vortex close to the bed. In addition, bed shear stress was considerably reduced in the scour hole compared to its value in the approach flow and was always smaller than the critical shear stress for the bed material. UDVPs have been used extensively to study the dynamics of particulate gravity currents (see Peakall et al., 2007 for a review) and the transport and settling characteristics of particulate matter (see Hunter et al., 2011 for a review).

4.5   OPTICAL TECHNIQUES: LASER DOPPLER ANEMOMETRY/VELOCIMETRY (LDA)

4.5.1   Introduction to LDA

Laser Doppler Anemometry (LDA) is an optical laser-based technique for measuring fluid flows (Durst *et al.*, 1981). It is an ideal technique for non-intrusive two and three component point measurements of flow velocities and turbulence distributions, providing single point measurements at very high spatial (<1 mm³) and temporal (>100 Hz) resolutions. It is thus well-suited to measuring high Reynolds number flows. Point measurements are obtained at the angular intersection of two high energy laser beams, which are positioned to cross at a measurement volume at a given focal length. Within this intersection, an interference fringe pattern of alternating light and dark planes is produced by constructive and destructive interference between the beams. Micron-sized seeding particles within the flow field scatter light as they move through this measurement volume and produce bursts of fluctuations in light intensity/amplitude as they pass through the bright and dark planes of the interference pattern. This burst pattern can be measured by the receiving optical arrays. The distance between the fringe pattern is known *a priori* and the frequency of light intensity amplitude modulation is the Doppler frequency, which is proportional to the flow velocity at the measurement point. These frequencies are captured by a receiving optical array and a photomultiplier converts the fluctuations to electrical signals which can then be processed and converted to give flow velocity information. Finally, as the temporal frequency of the obtained velocity signal is dependent on the arrival of seeding particles, the measurements are not spaced regularly in time. Wavelet analysis can be applied to convert between frequency-space and time-space, and to resample time-localisation information to obtain regularly-spaced temporal flow records for further analysis.

4.5.2   Strengths and weaknesses of LDA

Strengths:

- Non-intrusive;
- Three component velocity data;
- Very high frequency response (function of the flow rate but >100 Hz);
- Very high spatial resolution due to small sampling volume (<1 mm³);
- Extremely accurate.

Weaknesses:

- LDA systems (especially three component systems) are relatively expensive to purchase;
- Accuracy is highly dependent upon correct and accurate alignment of emitted and reflected beams, but laser alignment is non-trivial and time-consuming;
- LDA is capable of point measurements only and it is not really feasible to refocus on multiple sampling volumes owing to the nature of laser alignment;

- Line-of-sight is needed between the laser, the measurement volume and the photodetector;
- The performance of LDA is reduced by high levels of turbidity within the water column;
- LDA cannot measure shallow flows because the laser beams may intersect either the channel bed or the water surface;
- Focused laser light is potentially damaging to both flora and fauna.

4.5.3   Use of LDA in ecological studies

LDA has largely been used to measure velocities and turbulence spectra above either rigid (e.g., Nepf, 1999; Nezu & Onitsuka, 2001; White & Nepf 2007, 2008; Choi *et al.*, 2009; Liu *et al.*, 2010) or flexible (e.g., Nepf & Vivoni, 2000; Ghisalberti & Nepf, 2002) surrogate plant canopies or around live and model sessile animals (e.g., Monismith *et al.*, 1990; Butman *et al.*, 1994; Crimaldi *et al.*, 2007). In one study, Graba *et al.* (2010) employed three component LDA to measure the impact of biofilm growth on velocity profiles and turbulent properties at the crests and intervening interstice of two cobbles coated in biofilm. Data was collected for 240 s, resulting in the measurement of 10000 or 15000 velocity samples, which the authors reported were sufficient for good estimates of time-averaged velocities and turbulent shear stresses, respectively. In another study, Gambi *et al.* (1990) employed a three-beam, two-axis LDA to measure spatial distributions of 11–14 point vertical profiles of velocities above live eelgrass (*Zostera marina*) canopies of five different stem densities (400–1200 stems m^{-2}) and at three free-stream velocities (50–200 mm s^{-1}). Few details of the temporal parameters comprising the measurement protocol were provided.

4.6   OPTICAL TECHNIQUES: PARTICLE IMAGE VELOCIMETRY (PIV)

4.6.1   Introduction to PIV

Particle Image Velocimetry (PIV) relies upon illuminating a fluid, usually with a pulsed laser focused by a system of cylindrical and spherical lenses into a planar light sheet, and simultaneously capturing sequential images of lightweight particulate matter suspended within the fluid. PIV therefore does not directly measure the velocity of a fluid, but rather the motion of particles suspended within it. Thus, seeding particles must be selected to minimise lag between the motion of the fluid and that of the particles; i.e., seeding material should be as close to neutrally buoyant as possible. In addition, the diameter of seeding particles must be small relative to the resolution of the camera, but large enough that individual particles are imaged. Hollow glass spheres with a nominal diameter of 30–40 μm are commonly used. The displacement of the particles (in pixels) between images is computed using two-dimensional window-offset cross-correlation (Westerweel, 1997). Conversion from image (displacements measured in pixels) to physical space (velocities measured in m s^{-1}) is given by:

$$\mathbf{v} = \frac{\Delta x}{\Delta t}\left(\frac{L}{N}\right)$$

(4.20)

where $\mathbf{v}$ = two-dimensional velocity vector (m s^{-1}), Δx = particle displacement (pixels), Δt = time interval between images (s), L = reference length in physical space (m), N = number of pixels that image the reference length (pixels). Displacement fields (in pixels) are thus converted into two-dimensional velocities using the known time interval between image frame capture, obtained from a timing synchronization unit, and the measurement of a known length within the image. A calibration target comprising a double-sided multi-plane black stainless steel plate with white 10 mm-spaced dots, is used to establish the length scales prior to commencement of experiments.

Spurious particle displacements, whilst rare (commonly less than 7% of the total), are ubiquitous features of PIV output because of either background noise within images or the dispersal of particulate matter. Validation of each dataset is therefore essential. Four different kinds of validation techniques are commonly implemented:

1   a local median-filtering method, to identify displacement vectors that deviate by a prescribed amount in magnitude or direction, from adjacent vectors (Westerweel, 1994);
2   a cross-correlation Signal-to-Noise Ratio (SNR) validation, where the highest correlation peak is compared with the second one, and validated if the ratio is greater than a predefined value (commonly 1.2, Keane & Adrian, 1992);
3   a displacement range validation, which rejects vectors outside a certain velocity range;
4   a geometric validation, which rejects vectors within a certain predefined area. This is useful when solid surfaces are within the area of investigation.

A more detailed description of PIV and the quantification of its measurement uncertainty can be found in Raffel *et al.* (1998).

4.6.2   Strengths and weaknesses of PIV

Strengths:

- Non-intrusive;
- Two component velocity data in basic mode, three component data is possible through development of stereoscopic PIV systems and/or scanning systems;
- Very high frequency response (potentially >1000 Hz with modern high speed cameras);
- Very high spatial resolution (as small as the Kolmogorov microscale with modern high resolution cameras), limited only by the quality of the camera lens and pixel size/resolution of the camera. Spatial resolution is constantly improving with advances in both digital optics and computing;
- Capable of imaging flow fields even in extremely shallow water.

Weaknesses:

- They are expensive to purchase;
- Setup is not trivial. Significant care must be taken to correctly focus the camera(s) within the correct plane(s) in space;
- Requires large computing capacity;
- Accurate synchronization of laser/light source pulses with image capture is a crucial factor in the relative success of PIV;
- It is essentially an indirect measure of fluid flow. Particulate matter within the water column is tracked and velocities are computed by cross-correlating sequential images. If the motion of these particles does not adequately represent that of the flow (e.g. if the density of the particles is significantly different to that of the fluid), unreliable estimates of the flow field will result;
- It is an optical technique and therefore line-of-sight is needed between the light source and the measurement volume and between the camera and the measurement volume. The fluid must be relatively clear;
- Objects moving unpredictably within the imaged plane limit the success of the cross-correlation algorithm. High quality PIV algorithms are therefore presently being developed to dynamically mask impacted regions of the flow;
- Inconsistent lighting (e.g. caused by absorption or reflection of light from objects or flume walls) may degrade image quality and hence introduce errors into the cross-correlation computation;
- Errors may be introduced if the seeding material contains disparate grain sizes and light reflective/absorptive qualities;
- Focused laser light is potentially damaging to both flora and fauna. However, it is possible to employ PIV using alternative (lower energy) light sources.

4.6.3 Use of PIV in ecological studies

At the time of writing PIV had been employed in a small number of studies to measure the flows within plants and around sessile and mobile benthic animals. Tinoco & Cowen (2009) used two component PIV to capture the mean and turbulent flow field immediately downstream of, and at the channel margin of, a live *Myriophyllum spicatum* canopy. Nezu & Onitsuka (2001) and Schindler *et al.* (2004) used two component PIV to study spatio-temporal patterns of *u* and *v* and their turbulent fluctuations around the canopy and stems of stiff bronze and wooden cylinders, respectively. Stamhuis *et al.* (2002) document a series of studies conducted with two component PIV at the Biological Centre of the University of Groningen, Netherlands. Their work covers animal behaviour over a wide range of Reynolds numbers, e.g. the swimming of larval and adult copepods (*Temora longicornis*) (0.1 < Re < 100) (Van Duren *et al.*, 2003); burrow ventilation in a mudshrimp (*Callianassa subterranea*) (Re ~ 200) (Stamhuis & Videler, 1998a, 1998b, 1998c); and burst swimming in zebra danio larvae and adults (*Brachydanio rerio*) (Re = 400–3000) (Müller *et al.*, 2000). Moulin *et al.* (2007) used three component PIV to investigate the impact of a staggered array of slipper limpet (*Crepidula fornicata*) models on the hydrodynamics within the benthic boundary layer. They reported that at least 600 image pairs were required for adequate statistical convergence of time-averaged velocities, while at least 1000 image pairs were required for adequate statistical convergence of the covariance.

4.7 THERMAL TECHNIQUES: HOT FILM AND HOT WIRE ANEMOMETRY (HFA/HWA)

4.7.1 Introduction to HFA/HWA

Hot film and hot wire anemometers operate by relating forced convection of heat from a sensor surface to the intensity of flow or shear, with a temperature gradient set up between the heated surface and the fluid (Gust, 1988). Thus, a surface is electrically heated to a temperature above that of the ambient fluid and the electrical resistance of the surface is monitored as it cools in the flow. Hot film anemometers consist of a 0.1 µm-thick conductive platinum film coated on any of a 25–150 µm-diameter quartz fibre, a 50 µm-thick hollow glass tube, or a quartz cone, wedge, parabola, or hemisphere. Hot wire anemometers have a 1–2 mm-long, 3–10 µm-diameter platinum or tungsten wire extended between two arms. The response of hot film anemometers is faster than that of hot wire ones because the conductive element is thinner, even though hot film probes are thicker in total for sturdiness. A calibration relationship is used to relate the output current, voltage or temperature to the fluid velocity.

4.7.2 Strengths and weaknesses of HFA/HWA

Strengths:

- High frequency response (up to 400 kHz);
- Small sampling volume: <1 mm^3.

Weaknesses:

- They are not particularly sensitive to low velocities and cannot measure negative ones, making deployment within plant canopies or behind obstacles problematic;
- They are not suitable for measuring moving objects;
- Care must be taken to not touch biota with the hot surface of the probe, as this may cause trauma;
- They usually provide one or two component velocities, although some three component devices are available, at a single point and do not allow simultaneous measurements through the water column. Thus, in order to obtain time-averaged velocity profiles, either multiple probes need to be deployed simultaneously or multiple point measurements must be combined, increasing the total measurement time;
- Both methods are intrusive and the probe acts as an obstacle to the flow;
- Both HFAs and HWAs are very sensitive to fouling, with performance impaired within a few seconds, making them impractical for obtaining some time-averages;
- Because of their extreme sensitivity to turbidity and temperature, both methods require frequent recalibration. They typically need to be removed from the fluid

after a maximum of two velocity measurements and cleaned with a fine sable hair brush to remove air bubbles or any debris (e.g., Biggs *et al.*, 1998; Thomson *et al.*, 2004).

4.7.3   Use of HFA/HWA in ecological studies

A handful of studies have employed HFA and HWA to either measure at-a-point velocities or to measure profiles in and around flora and fauna. For example, Biggs *et al.* (1998) used a 2-mm-diameter rugged side-flow HFA to measure vertical distributions of velocity at four locations around cobbles coated with biofilms. When a 2-axis LDA could not measure velocities within a *Zostera marina* canopy, Gambi *et al.* (1990) used HFA to extend their 11 to 14 point velocity profiles into the canopy. Leonard & Luther (1995) used an array of HWA sensors spaced 10 mm apart to sample velocities within two marshes comprising stands of *Juncus roemerianus*, *Spartina alterniflora* and *Distichlis spicata* in Florida and Louisiana. Sand-Jensen & Mebus (1996) used three to five HWA probes to measure spatial distributions of velocities within and around patches of macrophytes (*Callitriche cophocarpa*, *Elodea Canadensis*, *Sparganium emersum*) in seven small, shallow Danish streams. HWA has also been used within and above stands of rigid wooden cylinders to investigate the impact of varying stand height on two component turbulent velocity profiles (Seraphin & Guyenne, 2008). Fiala-Médioni (1978a) used HFA to make *in situ* measurements of exhalent velocities from four species of ascidians (*Ascidia mentula*, *Phallusia mammillata*, *Styela plicata* and *Microcosmus sabatieri*). Finelli *et al.* (2002) and Thomson *et al.* (2004) used HFA to study *Simulium vittatum* larvae. Finelli *et al.* (2002) investigated the effects of velocity and particle concentration on the posture, flick rate, and ingestion rate, while Thomson *et al.* (2004) measured the effect of labral fan motion on flow patterns downstream of the larvae. Eckman & Nowell (1984) measured spatial distributions of boundary shear stress around individual tube mimics (3–6 mm-diameter rigid circular cylinders) (see Table 4.1 for additional examples).

4.8   DRAG MEASUREMENTS

4.8.1   Introduction to measuring drag

Measuring the form drag of objects within a flow field (e.g. vegetation) is important for understanding the distribution of velocities and resulting shear stresses. Fluid drag can be estimated using measurements of the shear stress acting over a given surface area. Ackerman & Hoover (2001) describe a number of different methods for estimating bed shear stress and drag forces (see Table 4.3). However, uncertainties in these estimates, particularly where there are strong local gradients in velocity or when near-bed measurements are difficult to obtain, may lead to significant errors in drag force measurement.

Tinoco & Cowen (2013) highlight the difficulties associated with direct or indirect measurement of drag forces due to: strong local gradients (due to flow blockage

Table 4.3a Direct techniques for measuring wall shear stress (adapted from Ackerman and Hoover, 2001).

Measurement technique	Location	Limitations and/or requirements of technique use	Reference(s)
Force balance techniques (measure the force exerted on the boundary to estimate bed shear stress)			
Floating-element balance (various types)	on/ beneath	Difficulties with gaps and alignment; tradeoff between element size and sensitivity; zero pressure gradient required	Winter (1977); Hanratty & Campbell (1983); Haritonidis (1989)
Surface coating techniques (measure changes in the surface properties of the material to estimate bed shear stress)			
Oil film interferometry Liquid crystal and pressure-sensitive films/paints	on	Require visualization for data acquisition; no calibration needed; surface may be fouled	Fernholz et al. (1996) Buttsworth et al. (2000)

Table 4.3b Indirect techniques for measuring wall shear stress (adapted from Ackerman and Hoover 2001).

Measurement technique	Location	Limitations and/or requirements of technique use	Reference(s)
Momentum balance techniques (relate the momentum change in the flow to the force exerted by the wetted perimeter to estimate the force applied)			
Pressure gradient – in channels	–	Difficult to apply when the wetted area is not constant	Haritonidis (1989)
Momentum thickness	–	Difficult to apply to complex geometries	Haritonidis (1989)
Wall similarity techniques (based on velocity or pressure measurements above and/or near the wall to estimate bed shear stress)			
(i) Velocity profile techniques (i.e., based on the law of the wall)			
Depth-slope product	above	Estimate of boundary shear stress. Assumes uniform flow	Chow (1959)
Near-bed velocity measurement or regression of the velocity profile	above	Sensitive to the shape of the velocity gradient; >5 observations are required in the logarithmic layer	Nowell & Jumars (1984); Wilcock (1996); Ackerman & Hoover (2001)
Near-bed Reynolds stresses, and energy-dissipation techniques	above	High spatial and/or temporal resolution are required in the viscous sublayer and the logarithmic layer, respectively	Dade et al. (2001)
(ii) Obstacle flow techniques (i.e., based on flow near the wall)			
Preston-tube	on	Probe must be oriented correctly; problematic in highly accelerating flows, detached boundary layers, or where law of wall absent; relatively high pressure needed	Preston (1954); Ackerman et al. (1994); Fernholz et al. (1996); Ackerman & Hoover (2001)

(Continued)

Table 4.3b (Continued).

Measurement technique	Location	Limitations and/or requirements of technique use	Reference(s)
Surface obstacle (Stanton tube, subsurface fence/ gate, block, step)	beneath	Probes can be fouled	Haritonidis (1989); Fernholz et al. (1996)
FST hemispheres	on	Provides an integrated shear stress, boundary conditions affected by the deployment platform	Statzner & Müller (1989); Frutiger & Schib (1993); Dittrich & Schmedtje (1995)
(iii) Mass/heat transfer techniques (i.e., relate mass/heat transfer to bed shear stress through calibration)			
Heat transfer techniques (hot wire, pulsed wire, hot film)	on	Requires frequent calibration; prone to fouling; some types can be used where flow reversals occur	Hanratty & Campbell (1983), Gust (1988); Fernholz et al. (1996)
Dissolution and liquid transfer Techniques	on	Dissolution rate affected by turbulent fluctuations; prone to fouling; material may affect surface conditions	Porter et al. (2000)
Polarographic (electrochemical) techniques	on/ beneath	Difficulties associated with the choice of electrochemical system and interactions with the fluid	Hanratty (1991)

by the object); the effects of unsteadiness (due to wave effects); motion or deformation of the object (e.g. the bending of plant stems and reconfiguration of foliage); the need to take measurements close to the boundary; and the complex interactions between multiple objects (e.g. flow through plant canopies). Most indirect approaches (Table 4.3b) rely on velocity measurements near to the boundary and may be difficult to implement in physical modeling with biota.

Conversely, embedded sensors can be used for direct measurements of fluid drag (e.g. Fathi-Maghadam & Kouwen, 1997; Armanini et al., 2005; Wienke & Oumeraci, 2005; Schoneboom et al., 2008) and arrays of sensors (e.g. Sand-Jensen, 2003; Schoneboom et al., 2010) or drag/shear plates (e.g. Callaghan et al., 2007; Tinoco & Cowen, 2009, 2013) may be used for measuring the combined impact of many individual objects (Table 4.3a).

4.8.2 Strengths and weaknesses of drag measurement

Indirect measurement:
Strengths:

• Can be obtained from standard flow measurement techniques.

Weaknesses:

• Problems of measuring velocities near the boundary and where gradients are steep and/or complex.

Direct measurement:
Strengths:

- Avoids problems where velocities cannot be directly measured by acoustic or optical techniques.

Weaknesses:

- The cost and complexity of devices to measure forces on individual objects.

4.8.3   Applications in ecological studies

Fathi-Maghadam & Kouwen (1997) designed a force balance apparatus to measure drag forces as large as 100 N acting upon individual pine and cedar saplings. Saplings were mounted inside a glass cone and attached to a table that was supported by four knife-edge frictionless legs in a Perspex box installed beneath the flume. The cone was laid flush with the flume bed. A load cell was then installed between the table and the Perspex box. Experiments were run at five flow depths ranging from 60 mm (partial submergence) to 300 mm (total submergence). Fathi-Maghadam & Kouwen (1997) concluded that for flexible vegetation, drag does not vary with the square of the approach velocity (as may be expected considering the classical drag equation), but instead varies linearly. They ascribed this difference to reductions in both plant projected (or momentum-absorbing) area and the drag coefficient as velocity increases.

Sand-Jensen (2003) mounted plant shoots and plastic leaves on a PVC plate mounted 0.25 m below the water surface. The plate was connected by thin metal supports to a platform that "hovered" on a second platform through which compressed air was blown out of small holes. This arrangement followed unsuccessful attempts to build a frictionless device using Teflon bearings. Forces acting on the test objects were measured using spring balances with various different measurement ranges. Results showed that i) increasing flexibility leads to greater reconfiguration and lower drag coefficients, ii) there is a greater decrease in drag coefficients for flexible plants compared to rigid plants with increasing flow velocities, and iii) bending of flexible plants results in shielding which significantly reduces drag.

Armanini *et al.* (2005) designed and built a drag plate capable of measuring three-component forces acting upon plant stems. Individual *Salix alba* (white willow) trees were firmly clamped onto a movable upper steel plate that was bound to a lower static steel plate by four deformable aluminium foils. Deformation was monitored by eight strain gauges attached to the foils, and suitably coupled to form four Wheatstone bridges, allowing the measurement of the three components of the force and the bending moment. The entire unit was waterproofed and then buried beneath the floor of the flume with sand, gravel and concrete. Armanini *et al.* (2005) concluded that partially submerged willows behaved as rigid cylinders because of the limited number of leaves and stems that could reconfigure, with the drag coefficient remaining constant and the drag force increasing with the square of flow velocity. However, when fully submerged, the thinner branches and leaves that are more numerous near the top of the canopy tended to reconfigure, reducing the projected area-drag coefficient product, AC_D, and causing the drag force to instead increase linearly with the flow velocity. In addition, removing foliage reduced drag forces by up to 40% relative to foliated conditions.

Callaghan *et al.* (2007) designed a Submersible Drag Gauge (SDG) to measure one-component (usually longitudinal or streamwise) drag forces acting on partially or fully submerged vegetation, sediment, or other objects. The SDG consists of a moveable low friction shear plate of two alternative designs (to cater for light or heavy loads, respectively) installed within a stationary, rigid housing supporting a load cell calibrated in compression. Callaghan *et al.* (2007) validated the SDG by comparing their results obtained with a stiff cylinder to literature values. The SDG has been employed by Statzner *et al.* (2006) and Callaghan *et al.* (2007) to measure drag forces acting upon a stiff cylinder and live and artificial plants (Anacharis; *Egeria densa*). The setup of Statzner *et al.* (2006) is notable because they also obtained concurrent images of the 3-D reconfiguration of the plant using angled mirrors to image the frontal area. This permitted them to observe how changes to the frontal projection area and lateral projection length of a given plant occurred for various flow rates and how the drag on a given plant varied following random reconfiguration of the plant. Such reconfiguration was more marked than had previously been observed in stiffer (but still flexible) tree crowns by Rudnicki *et al.* (2004). Because of the greater streamlining permitted by the more flexible live plants, drag forces increased linearly with velocity for the live vegetation (exponent 0.97), but non-linearly for the other two cases (exponent ~1.53 in both cases). In addition, objects with larger projected area experienced more drag, as expected from classical rigid body theory.

Boller & Carrington (2007) measured the drag acting on macroalgae using a custom-built force platform that measured forces in three dimensions. Macroalgae were attached to a post which was connected to a three-axis ceramic force transducer measuring at 100 Hz for 10 s. Results from the force measurements were used to relate drag coefficients to the different structural properties of the macroalgae.

Fonseca *et al.* (2007) used force transducers connected to seagrass shoots to measure flow-induced forces associated with different spatial distributions of shoots. Measurements were processed using a Vishay Strain Indicator and data were recorded at 50 Hz for 250 s. Results showed that for unidirectional flow, forces were reduced when shoots were arranged in rows rather than randomly distributed. For oscillatory flow, results were inconclusive.

James *et al.* (2008) built a pivoting frame across their flume and initially attached a line to one end of it, which ran through a pulley and was connected to a weight. To the other end of the pivot, they sequentially affixed a 5 mm-diameter circular rod, a 5 mm square rod, a 25 mm square rod, and real and artificial *Phragmites australis* (2×) and *Typha capensis* (1×) stems. Application of moment equilibrium enabled the drag force to be calculated. However, because this setup was found to be inaccurate for measuring small forces, the weighted line was replaced by a system of four load cells positioned against a transverse bar 0.433 m below the pivot and connected to a static strain indicator. James *et al.* (2008) found that i) stem shape affected the value of the projected area-drag coefficient product, AC_D: for Re < 1000, AC_D for square rods varied more strongly with Re than that for round rods. For 1000 < Re < 10000, AC_D values for square rods were ~30–50% higher than for round rods; ii) dense foliage resulted in larger AC_D values than scanty foliage; iii) AC_D values obtained for single plants approximated those obtained for stands of plants; and iv) if proper account is taken of streamlining, the variation of AC_D with Re is consistent with that observed for rigid rods, implying that the variability that James *et al.* (2008) observed in AC_D is associated more with plant reconfiguration, and hence with A, rather than with C_D itself.

In a similar technique to that employed by Armanini *et al.* (2005), Schoneboom *et al.* (2008, 2010) developed a drag force measurement system (DFS). Their system consisted of a vegetation element mounted on a movable aluminium plate that was connected to a second fixed plate by a 140 mm long, 20 mm wide, and 3 mm thick stainless steel beam. When subjected to drag, the plant acted as a cantilever in bending and generated compression strains that were monitored by eight strain gauges at two vertical positions on the steel beam. The strain gauges were configured as two full Wheatstone bridge configurations, enabling drag force measurements with a high accuracy (standard error ±0.02 N) and temporal resolution (1613 Hz). In initial tests with the DFS mounted upside down above a race-track flume, Schoneboom *et al.* (2008) validated measurements using tests on rigid cylinders and then performed experiments with three types of natural and artificial vegetation: natural and plastic flexible poplar (*Populus nigra*) and willow (*Salix alba*) branches, artificial sedges (*Carex* sp.) and rigid steel and plastic cylinders with diameters of 10 mm. They found that the precision of the drag force measured on rigid cylinders was reduced by flow-induced vibrations at higher velocities, which were significant for the plastic cylinder but limited for the steel one. As reported by other studies, drag forces on rigid elements increased with the square of the velocity, while those on flexible elements increased linearly with velocity. Schoneboom *et al.* (2010) employed an array of 10 DFS sensors sampling at 200 Hz to study the impact of stem arrangement (i.e. staggered versus inline arrangements) and areal density (number of plants per unit area) of artificial poplars on drag forces. They concluded: i) drag increases with the square of velocity at low flows because of the limited deformation of stems and leaves, but then increases linearly with velocity at faster flows; ii) drag forces were consistently larger (by approximately 1.23 times) for staggered arrays rather than for inline arrays; iii) drag forces measured on inline arrays were similar to those measured on single elements; and iv) form drag on vegetal elements contributed 75%–95% of the total shear stress within the flume.

Xavier *et al.* (2010) and Wilson *et al.* (2010) adopted an inverse approach to study the drag acting upon the stems and leaves of 22 fully submerged trees of the genera *Salix*, *Alnus* and *Populus*. Plants were suspended upside-down below a carriage and dragged at rates of up to 6 m s^{-1} through a large calm-water towing tank. A dynamometer was used to measure the drag forces acting on both foliated and defoliated samples at various velocities. Experimental results identified a critical velocity at which flexing behaviour became important and found that foliage contributed between 24.4% and 54.8% of the total drag.

Tinoco & Cowen (2013) developed a non-intrusive drag measurement device that is based on isolating a section of the flume bed and fitting a plate that can move on frictionless rails. The force applied to the moving plate can then be directly measured. The size of drag plate can be chosen to enable spatial averaging over multiple elements such as patches of macrophytes. They validated the measurements of the drag plate using simple structures and showed that it could be applied to rigid or flexible vegetation fixed to the flume bed.

4.9 CONCLUSIONS

The flow measurement literature shows considerable variability between studies and authors in experimental design and/or reporting. In the closure to their paper "Turbulence measurements with Acoustic Doppler Velocimeters", García *et al.* (2007) present general guidelines on how to perform velocity measurements capable of quantifying turbulence in fluid flows. An edited version of these guidelines is reproduced here, to encourage the standardised reporting of results and hence to increase comparability between studies:

1   Define the objectives of the study (e.g., characterization of mean values, turbulent kinetic energy, Reynolds stresses, turbulence length, space and time scales, etc...);
2   Define the flow regions where flow turbulence characterisation is intended (e.g., near-bed, near free surface, around objects, etc...);
3   Determine the sampling duration for each flow zone depending on the objective of the study as defined in step 1 (section 4.1.1);
4   Determine the optimal sampling frequency in each region of the flow required to characterize flow turbulence parameters (i.e., $f > 20\overline{u_i}/L_e$, where L_e is the energy containing eddy length-scale);
5   Perform a pilot study, including sensitivity analysis, to enable the definition of the optimum instrument type and/or sampling configuration (i.e. velocity ranges, size of the measuring volume, etc...) for each region of the flow and the selected sampling frequency. The optimum configuration provides the best signal quality for the observed flow conditions. For instruments that require neutrally buoyant seeding particles to enable flow measurement, the seeding density should be sufficient to minimise measurement errors (e.g. spikes in ADV data or poor temporal resolution in LDA and PIV data);
6   Perform error-checking on the velocity time series. Remove spikes and replace them in cases where the object of the study entails the analysis of the temporal signal correlation; and
7   Define confidence intervals about each of the computed turbulence parameters (e.g. García *et al.*, 2006).

Although some of these points are specific to ADVs, many are pertinent for other techniques as well. Few of the papers cited in this chapter address all of these points. Most studies fail at points 3 to 5. Identifying the sampling duration, sampling frequency, velocity range and size of measuring volume is potentially an iterative process, but can be informed through consideration of section 4.1. The sampling volume should ideally be as small as possible, in order to resolve the smallest energy-containing eddies, limit spatial averaging and reduce the impact of velocity gradients and shearing within the sampling volume. For their specific case, Crimaldi *et al.* (2007) estimated the dissipation (Kolmogorov) length scale as 0.1 mm, which is considerably smaller than the dimensions of the sampling volume of ADVs, ADVPs and UDVPs and is of similar order to the dimensions of HFAs and HWAs, LDAs and modern PIV systems. Thus, the smallest eddies are unlikely to be sampled with present technologies. If the sampling duration employed in the pilot study advocated in point 5 is longer than 8 times a reasonable estimate of the integral timescale and the sampling frequency is the largest that available equipment will allow, results obtained from the pilot study may be used to estimate the optimum sampling duration and sampling frequency. Subsequent analysis of data, investigating convergence rates of the variables of interest and turbulent power spectra should then yield appropriate values that can be employed in the final experiments and also permit estimation of the Kolmogorov length scale for the specific conditions encountered in the pilot study. At present, many studies merely report that velocities are measured with a certain technique, with little elaboration on either the sampling frequency and/or sampling duration. Of those studies that do report the sampling frequency and duration, an even smaller fraction report upon the extent to which time-averaged velocities and turbulent quantities have converged.

A crucial factor in any experimental design is the identification and mitigation against or elimination of errors (point 6). However, few studies performed with acoustic techniques mention post-processing the raw time-series prior to interpretation and, of those, very few specify the post-processing steps that were undertaken. For example, few studies using ADVs mention eliminating data if the correlation coefficient, R^2, is less than 0.6 and if the signal-to-noise ratio is less than ~15 dB. Finally, (point 7), very few studies quantify the uncertainty associated with both time-averaged velocities and turbulent quantities. In one exception, Folkard & Gascoigne (2009) calculated the integral time scale (although they did not provide exact values, merely stating that it was ~0.045–4.5 s) and estimated the uncertainties in their ADV measurements (sampled at 25 Hz for 90 s), resulting in estimates of 2–3% error in the mean and 15–16% in TKE and Reynolds stress. Similarly, Lassen et al. (2006) computed a typical integral time scale of 0.6 s and estimated errors of 5% in the mean and 14–34% in the velocity variance and covariance. Monismith et al. (1990) estimated that their time-averaged velocities obtained with LDA were accurate to within ± 2 mm s^{-1} and that $\overline{u_i'^2}$ values had an uncertainty of ~13%.

Any researcher interested in making meaningful measurements of the interactions between flow and biota must consider very carefully the points made in this chapter when designing their experiments.

Organism specific considerations

Chapter 5

Biofilms

P. Saunders, F.Y. Moulin, O. Eiff, M. Rogerson &
H.M. Pedley

5.1 INTRODUCTION

Biofilms thrive in a wide range of both natural and man-made environments, the only limiting factor being the presence of adequate water (Sutherland, 2001). They are found in aquatic and soil environments, on plants, pipes and filtration systems and even on the tissues of animals and humans (Bradding *et al.*, 1995; Wingender *et al.*, 1999; Flemming & Wingender, 2001; Sutherland, 2001). At the simplest level, biofilms can be considered as communities of microorganisms attached to a surface (O'Toole *et al.*, 2000). Depending on their location, they are given different names. In river, lake and estuarine environments, the "periphyton" (or periphytic biofilm) groups all biofilms growing in the vicinity of macrophytes, including the "epiphyton" (or epiphytic biofilm) attached to the stems and leaves of macrophytes. Biofilms on the bed are generally sub-divided based on the kind of substrate they are attached to, for instance, the "epilithon" or epilithic biofilm grows on the sediment whereas the "epixylon" or epixylic biofilm grows on wood. However, other names can be found in the literature, for example, "sediment biofilms" for growth on fine-grained (<2 mm diameter) bed sediments (i.e. on clays, silts or sands) and "microphytobenthos" for photosynthetically active (autotrophic) biofilms growing on marine mudflats.

Biofilms have the capacity to change the physical and chemical characteristics of the aqueous environment in which they reside. These effects include altering flow dynamics, total suspended sediment concentrations, transient storage volumes, pH and elemental concentrations. Conversely, the physical and chemical nature of the aqueous environment can cause substantial changes in biofilm mass, architecture and species diversity, which in turn can further alter the aqueous environment. A complex relationship of feedback mechanisms therefore exists between biofilms and the water surrounding them. It is essential that hydraulic experiments take these factors into account to avoid the production of "contaminated" data and thus maximise the reliability of any conclusions drawn.

5.2 BIOFILMS IN NATURAL HYDRO-ECOSYSTEMS

5.2.1 Composition and structure of biofilms

A biofilm consists of microorganisms such as bacteria, fungi, protozoa and algae, as well as biogenic and inorganic particles, multivalent cations and dissolved compounds contained within a matrix of Extracellular Polymeric Substances (EPS) (Wingender *et al.*, 1999;

Vu *et al.*, 2009). The EPS generally accounts for between 50 and 90% of the total organic matter of a biofilm (Wingender *et al.*, 1999) and in the majority of cases the dried mass of a biofilm is over 90% EPS, with the microorganisms themselves making up less than 10% (Flemming & Wingender, 2010). Microorganisms are able to alter the EPS matrix after it has been secreted from the cell. For example, the shape and electromagnetic charge can be changed by the addition of substituents to the polysaccharides and programmed cell death can lead to the creation of new pores and channels within the matrix (Flemming, 2011). The nature of the EPS matrix may also be altered by higher level organisms such as larvae, protozoa and snails that graze on it, resulting in the selective removal of EPS and EPS producing organisms. It has also been noted that the distribution of cells and EPS in a biofilm varies greatly; large areas of biofilm may consist mainly of EPS and actually be devoid of microbial cells (Wolfaardt *et al.*, 1999). The EPS matrix is highly intricate and complex, even for biofilms formed in the same environment; this complexity and the difficulties in analysing it has led to it being described as the "dark matter" of biofilms (Sutherland, 2001; Flemming *et al.*, 2007).

The development of a biofilm represents a lifestyle change for the organisms where they switch from a unicellular planktonic existence to a multicellular static state, with further growth of the microbial mass creating complex structured communities (Lemon *et al.*, 2008). The stages of biofilm formation start with the initial surface attachment followed by a monolayer formation, the development of multilayered microcolonies, the production of an extracellular matrix, and finally the establishment of a mature biofilm with a three dimensional architecture (O'Toole *et al.*, 2000). This process is shown schematically in Figure 5.1.

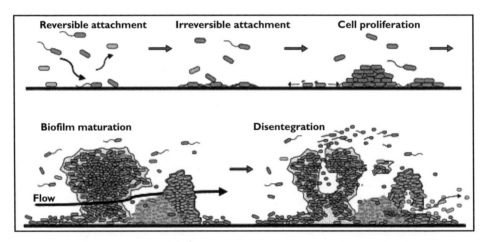

Figure 5.1 The biofilm development cycle. This first step which is most often reversible, may require active swimming motility or may just be caused by random contacts. In a second phase the attachment is fixed by adherence of the cells to the substratum through surface appendages or production of EPS. The third phase represents growth of the attached cells into microcolonies based on the available nutrients. In a hydrodynamic environment the development of the microcolonies depends on cell–cell binding interactions. The fourth stage is often referred to as the mature biofilm. At this point the biofilm structure with its distribution of biomass and the presence of water-filled voids illustrates the heterogeneity of the biofilm and the rigid properties of the developed structure. Finally, at some point the biofilm may partially dissolve, releasing cells that may move away to other locations where a new cycle can begin (Molin & Tolker-Nielsen, 2003).

Biofilm formation has been described as occurring in three stages (see Jackson *et al.*, 2001; Martiny *et al.*, 2003), adhesion, biofilm growth and biofilm maturation. During adhesion the initial selection of microorganisms is mainly controlled by the recruitment of planktonic species from the water (Jackson *et al.*, 2001; Lyautey *et al.*, 2005). Diversity is high at this stage as there is minimal selection pressure due to the low number of competing organisms. As the number of cells increase within the biofilm, the nutrient levels in the water become increasingly important, competition for resources intensifies, and less competitive species are eliminated (Besemer *et al.*, 2007), resulting in a reduction in diversity. Then, as the biofilm matures (often referred to as a fourth stage) it develops a complex three dimensional architecture with multilayered microcolonies held within a matrix of EPS. This generates a variety of microhabitats allowing diversity to increase again.

Heterotrophic organisms are usually found in close proximity to the phototrophic community as they are able to thrive on the organic exudates the phototrophs produce. The presence of extracellular polysaccharides within the biofilm matrix allows photosynthetic organisms to thrive in deeper parts of the biofilm than might be expected (Flemming, 2011). Extracellular polysaccharides have a slightly different refractive index than water which results in a forward scattering of photons rather than a back scattering reflection. Light will therefore tend to propagate through a biofilm rather than being reflected from it. Also, the gelatinous nature of the extracellular polysaccharides enables light to scatter photons within deeper layers and therefore increase the chances of absorption by photosynthesising organisms deep in the biofilm (Flemming, 2011).

5.2.2 Biofilms in natural hydro-ecosystems

In a freshwater setting, biofilms are generally dominated by diatoms and cyanobacteria which represent the major primary producers. Other constituents include fungi, protozoa, filamentous green algae and heterotrophic bacteria such as sulphate reducers, anaerobes and methanogenic bacteria (Shiraishi *et al.*, 2008; Dürr & Thomason, 2009). In marine environments, biofilms contain different species of diatoms, bacteria, Archaea and flagellates; the amounts of other microorganisms such as sarcodines, ciliates and fungi are low compared to freshwater biofilms (Dürr & Thomason, 2009).

For sediment biofilms in rivers, lakes or mudflats, the biofilm structure near the surface is less complex than for epilithic biofilms grown on coarser substrates (Underwood *et al.*, 2005), but the presence of EPS generates adhesion between sediment grains that leads to a large increase in the critical shear stress required for bed erosion (Righetti & Lucarelli, 2010). For autotrophic biofilms in intertidal mudflats, the diatoms that dominate the photosynthetic species migrate vertically within the sediment bed in response to diurnal and tidal cycles (Kingston & Gough, 2009), resulting in fluctuations in bed sediment erodibility at similar timescales.

In flood-regulated systems, the succession of bacterial and algal composition is reflected in periphyton biomass variations that generally exhibit the idealised behaviour schematised in the benthic algal accrual curve of Biggs (1996). This curve shows that after colonisation and the initial exponential growth of biomass (measured as Dried Matter DM, Ash-Free Dry Matter AFDM, or Chlorophyll a for autotrophic biofilms), a maximum value is reached prior to a slight reduction towards a mature equilibrium value. This equilibrium is disturbed by sloughing or the removal of large sections of biofilm by the flow. Sloughing can either be "exogenic", relating to a catastrophic detachment driven by a flood or turbidity (see Boulêtreau *et al.*, 2010) or "autogenic", triggered by a temperature-induced bacterial community explosion (see Boulêtreau *et al.*, 2006). In non flood-driven systems (e.g. lakes; Jackson *et al.*, 2001), biofilms exhibit non-synchronous growth and thus a quasi-random spatial pattern of biofilms of varying maturity results, preventing the formation of the succession identified by Biggs (1996).

5.2.3   Factors responsible for biofilm selection and growth

Developing biofilms constantly undergo compositional and structural changes in response to both environmental (allogenic) and "self determining" (autogenic) factors. Generally, for epilithic biofilms the initial colonisation of a bare substrate is by diatoms which subsequently alter the microenvironment making it suitable for other organisms (Stevenson *et al.*, 1983). However, in more extreme environments, such as high or low light intensities, initial colonisation tends to be dominated by green algae or heterotrophic bacteria, respectively (Roeselers *et al.*, 2007). Early researchers concerned with developing autotrophic periphytic biofilm growth models identified and partially quantified the major external factors responsible for biofilm selection and growth (e.g., Stevenson, 1983, 1996; Peterson, 1996). These are nutrient concentration, temperature, illumination (affected by shading and turbidity), shear force and grazing by higher level organisms. Of these factors, shear force (fluid/air flow or particle to particle attrition) is generally considered to have the strongest influence on biofilm formation in hydrodynamic environments (Liu & Tay, 2002).

A number of studies have shown that diversity during biofilm development is directly related to shear stress (see Cloete *et al.*, 2003; Rickard *et al.*, 2003; Rickard *et al.*, 2004). Rochex *et al.*, (2008) found that species diversity is highest throughout biofilm development in very low shear stress conditions, with moderate levels of diversity occurring when these stresses increase slightly. As shear stresses increase, temporal variations in biofilm diversity were noted: intermediate levels cause a reduction in diversity in week 2 but an increase in week 3. At high shear stresses, diversity also decreases in week 2 and subsequently remains low. Structurally, strong shear generates thin dense biofilms whilst more gentle flows result in thicker and less dense structures (Figure 5.2). When the shear stress is greater than the internal strength of the biofilm matrix, detachment from the substrate will occur (Pei-shi *et al.*, 2008). In addition to shear stress, sloughing of biofilm material may result from the aging of deeper algal components, heterotrophic degradation and its associated gas bubble production (Boulêtreau *et al.*, 2006).

Flow velocity also affects biofilms indirectly, via regulating the supply of dissolved sediment to the EPS surface. Where biofilms are mineralising, increasing shear stress accelerates the process (Hammer *et al.*, 2008) potentially resulting in exotic precipitate morphologies which themselves locally accelerate flow (Pedley and Rogerson, 2010). Even in the absence of mineralisation, promoting ion supply to a biofilm enhances the supply of nutrients (e.g. phosphate or oxygen) and removal of potentially limiting metabolites (e.g. carbon dioxide). Enhanced supply of cations to the EPS may also affect (positively or negatively) the physical stability of biofilms due to changing binding behaviours. Finally, all these effects may result in a change in the species composition of the biofilm, changing the physical structure of the community.

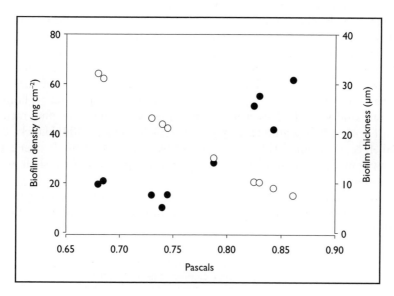

Figure 5.2 Effects of shear stress on biofilm density and thickness in steady state fluidised bed reactors (•): density; (o): thickness (adapted from Liu & Tay, 2002; Chang *et al.*, 1991).

Refinement of the earliest models (specifically, parameterisation of the nutrient-limiting parameters in the exponential growth term of the logistic biofilm growth model of Uehlinger *et al.*, 1996), led to a need for field surveys and laboratory experiments to better quantify the effect of nutrient concentration on biofilm composition and biomass in different aquatic systems (see, for example, Bowes *et al.*, 2007 for rivers and streams; Ventura *et al.*, 2008 for lakes; Nilsson & Sundbäck, 1991 for microphytobenthos in tidal zones). Bowes *et al.* (2007) found that although increasing the Soluble Reactive Phosphorus (SRP) concentration above 109 µg l⁻¹ had no impact upon algal biomass in the River Frome, Dorset, reducing Soluble Reactive Phosphorus (SRP) concentration below ~90 µg l⁻¹ caused it to decline, with a 60% reduction at <40 µg l⁻¹. Nilsson & Sundbäck (1991) added water enriched with sodium nitrate (NaNO₃) (~77 µ mols of inorganic nitrogen, N) and K₂HPO₃ (~3.2 µ mols of inorganic phosphorus, P) to three aquaria to simulate nutrient levels and N:P ratios found in shallow coastal areas influenced by sewage and river effluents. They found that in the first week of their experiment, the primary productivity of biofilms (measured in mg C m⁻² day⁻¹) increased significantly and after three weeks, it was eight times the value observed in controls. A similar trend occurred for Chlorophyll a. Thus, the biomass of microalgae overall was 2–3 times higher in the aquaria enriched with N and P.

5.2.4 Thermal and photocycle effects

For a biofilm developed in the laboratory to function as it would in its natural environment, its physical environment must be approximated adequately and key to this are control of photosynthetically active radiation and temperature. The diversity of microorganisms held in a laboratory-grown biofilm exposed to specific light and temperature conditions is strongly dependent on species-specific growth rates and any species within the biofilm complex can only acclimatise to the imposed environment within their individual genetically-determined limits (Defew *et al.*, 2004).

Temperature control is challenging, but recent experiments have shown that for small (20 × 8 × 2.5 cm) flume systems, temperature can be controlled with a standard deviation (σ) of between 0.2 and 0.5°C (Table 5.1) even over periods of ~25–30 days.

Temperature variations have been shown to have a major impact on both microbial diversity and growth rates from photosynthesis and heterotrophic metabolism (Blanchard *et al.*, 1996; Watermann *et al.*, 1999; Defew *et al.*, 2004; Hancke & Glud, 2004; Salleh & McMinn, 2011). Figure 5.3 shows the relationship between a standardised maximum photosynthetic rate (P_{Max}) and temperature for a shallow marine benthic biofilm.

The association between P_{Max} and growth rate is well demonstrated by comparing the optimum growth rate as a function of temperature (Figure 5.4) found by Butterwick *et al.* (2005).

Table 5.1 Temperature control in flow-through microcosm experiments (Saunders, 2012).

Experiment	Mean temperature	1σ (°C)
Microcosm 1	12.2	0.2
Microcosm 2	14.3	0.2
Microcosm 3	16.3	0.2
Microcosm 4	18.3	0.2
Microcosm 5	20.6	0.5
Control 1	12.1	0.5
Control 2	14.3	0.2

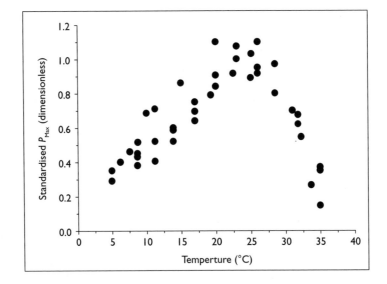

Figure 5.3 Standardised P_{Max} values for a diatom benthic biofilm (Blanchard *et al.*, 1997).

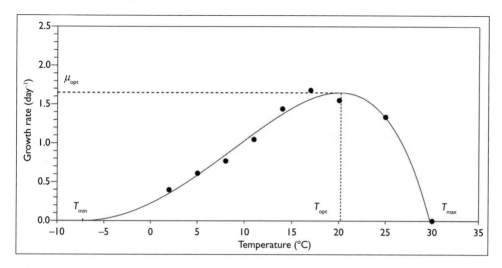

Figure 5.4 Optimum growth rates (μ_{opt}) ($T_{min, opt, max}$) as a function of temperature for *Asterialla formosa* (Butterwick *et al.*, 2005).

P_{max} values are species dependent, so the ability of any one species to outcompete and increase its percentage of the total biofilm composition varies with temperature. In diatom/cyanobacterial biofilms diatoms are the dominant organism at 10°C but at 25°C filamentous cyanobacteria dominate (Watermann *et al.*, 1999). After 14 days at 10°C, laboratory-grown biofilms appear to maintain an assemblage that closely matches their natural counterparts (Defew *et al.*, 2004), while at 18°C species composition is markedly different from the natural assemblage. At 26°C, diatoms quickly dominate, forming a thick biofilm. However, this larger biomass creates a high nutrient demand, the resultant reduction of nutrient availability combined with the physiological stresses on diatoms at high temperatures subsequently allow cyanobacteria to out compete diatoms (Defew *et al.*, 2004).

Practically, controlling the photocycle is straightforward and can be achieved using hydroponic lamps. The ideal light irradiance for cyanobacteria is between 15 and 150 µEinstein m^{-2} s^{-1} whilst for purple bacteria it is 5–10 µEinstein m^{-2} s^{-1} (Konhauser, 2007). Biofilms respond in a number of ways to variations in light intensity; studies of biofilm growth in light and dark conditions have shown that those grown in light have greater thickness, species diversity, algal density and biomass than those grown in the dark (Romani & Sabater, 1999; Sekar *et al.*, 2002). The differences do not develop immediately; during the first week of biofilm colonisation there is little difference between light and dark-grown colonies (Romani & Sabater, 1999; Sekar *et al.*, 2002). Differences in thickness stabilise after about 10 days but those in biomass continue to increase (Figure 5.5).

Of course, over the vast majority of Earth's surface, light intensities vary diurnally. Rogerson *et al.* (2010) performed mesocosm experiments in which they studied the impact of photosynthetic components of biofilms on bulk water pH. The diurnal changes in pH (Figure 5.6) were driven by an 18 hr to 6 hr day/night cycle. Lower pH

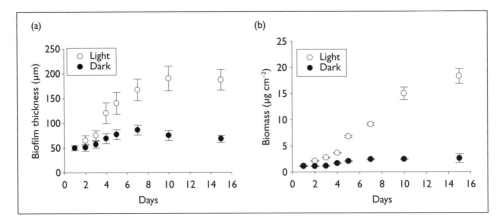

Figure 5.5 Thickness (a) and biomass (b) of biofilms grown under light and dark conditions for 15 days (Sekar *et al.*, 2002).

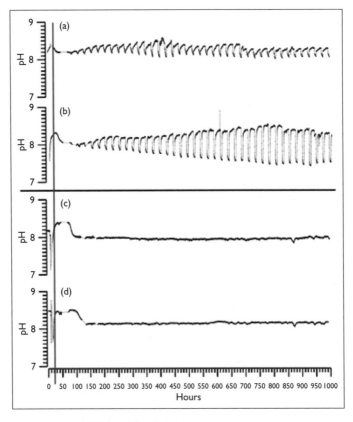

Figure 5.6 Amplitude of day/night variations in pH in mesocosms colonised with biofilm (a), (b) and in abiotic mesocosms (c), (d). Vertical line adjacent to X-axis indicates start of experiment. The low pH variations of the first 10 days represent the buffering capacity of the water rather than low microbial activity (Rogerson *et al.*, 2010).

values were observed during the 6 hour dark period in mesocosms colonised with biofilms than during the 18 hour light period clearly showing the impact of biofilm photosynthesis and respiration on solution pH in flume environments.

5.2.5   Spatial heterogeneity of biofilms

The heterogeneous nature of the structure of biofilms has been confirmed by Confocal Laser Scanning Microscopy (CLSM) (Stoodley *et al.*, 1999). Examination of both single and multi-taxa natural biofilms has revealed an architecture where slime-enclosed microcolonies are distributed between open cell-free water channels that penetrate all areas (Costerton, 1995). These water channels have been observed to allow both convective water flow and the passage of 0.3 μm polystyrene beads (Costerton, 1995). While patchiness has long been recognised, its significance has only recently been considered. Different mechanisms have been found to be responsible for patch development. First, the colonisation process itself is strongly dependent on local flow conditions, and therefore yields specific patterns at the substrate scale (as documented in recirculating flumes covered with very regular substrates by Labiod *et al.*, 2007 and Boulêtreau *et al.*, 2010). Second, interactions between flood events or storms of different magnitudes and biofilm development lead to spatial variability in age and sloughing resistance on different substrates. Third, detachment by sediment abrasion is highly dependent on local flow conditions and promotes patchiness. Indeed, field sampling protocols for biofilm characterisation minimise the effects of patchiness by either increasing the number of samples or the sampling area, or by substituting glass plates for natural substrates (although Nielsen *et al.*, 1984 identified significant differences between biofilms grown on such idealised substrates and those grown on natural beds).

More recently, spatial heterogeneities at the riverbed or mudflat scale, associated with mesoscale and macroscale bedforms (e.g. ripples, dunes, bars, etc), have attracted attention because of their ability to modify near-bed flow characteristics and hence the type of biofilm growing on different parts of the bed. However, heterogeneity is also known in relatively homogeneous environments ("patchiness"), even in relatively motile organisms (Murray & Alve, 2000), and its dynamics and origin remain poorly understood. Directly relating heterogeneity to flow characteristics should therefore be approached with caution. Besemer *et al.* (2009a) used artificial bedforms installed in a streamside flume to identify linkages between spatial heterogeneity of bed topography and biofilm composition and biomass. Community function, in terms of carbon uptake from injected glucose and dissolved organic carbon, is also enhanced by bed topography (Singer *et al.*, 2010). This is ascribed to increased heterogeneity in local flow conditions around bedforms that promote the growth of biofilm communities (measured by a global biodiversity index for the whole flume). Numerical modelling of step injections of conservative tracers by Bottacin-Busolin *et al.* (2009), approximating that employed by Singer *et al.* (2010), showed that nutrient retention in mature biofilms was roughly equal to retention in the porous bed. The authors of these three recent studies all recognised that the modification of local flow conditions by bedforms is directly responsible for the growth of different biofilm communities, and that this is the primary cause of all the differences observed (i.e., in biomass and microbiological biodiversity). This highlights the importance of local flow-biofilm interactions for understanding and predicting what happens at field scale. It is likely that the mechanism is at least partially the influence of shear stress on the supply of dissolved materials (Hammer *et al.*, 2008) but this link remains to be demonstrated empirically.

In addition to heterogeneity in morphological and species composition there are also variations in the EPS matrix that binds the community of microorganisms. The amount and composition of EPS and hence the structure of the biofilm matrix, depends on the type of microorganisms, the age of the biofilm and environmental conditions such as shear forces, temperature and nutrient availability (Sutherland, 2001; Vu *et al.*, 2009; Flemming & Wingender, 2010). As the above factors can change with time, the exact nature of this material within any biofilm varies in both space and time (Flemming, 2011). Given the ability of EPS to chelate divalent cations from the bulk water this is significant for flume experiments where constant water chemistry is required (see section 5.3.1 for discussion on chelation by EPS).

5.3   IMPACT OF BIOFILMS ON THE PHYSICO-CHEMICAL ENVIRONMENT

5.3.1   The impact of biofilms on biogeochemical fluxes and ecosystem function

Microbial biofilms play an important role in the complex interactions between the hydro-biogeochemical environment and ecosystem function, both through their metabolic processes and their secretions such as EPS. Laboratory (mesocosms) and field experiments have been used to quantify the impact of biofilms on the biogeochemistry of the whole hydro-ecosystem in particular nutrient uptake and transient storage, including work on dissolved organic carbon (Cazelles *et al.*, 1991; Battin *et al.*, 1999; Augspurger & Kusel, 2010), dissolved organic nitrogen (Kim *et al.*, 1990; Laima *et al.*, 2002; Arnon *et al.*, 2007; Wagenschein & Rode, 2008), and phosphorus (see Dodds, 2003; Horner *et al.*, 1990; Buzzeli *et al.*, 2000; Noe *et al.*, 2002; Gainswin *et al.*, 2006a, 2006b). Recently, location-specific quantitative studies have been undertaken for natural (e.g. Karathanasis *et al.*, 2003; Tao *et al.*, 2006, 2007a, 2007b; Li *et al.*, 2010) and constructed (Mitsch *et al.*, 1995) wetlands. These results have permitted the parameterisation of reactor-type models of the system of interest (see Cazelles *et al.*, 1991; Buzzeli *et al.*, 2000; McBride & Tanner, 2000; Doyle & Stanley, 2006).

The ability of biofilms to take up, store and/or recycle pollutants and poisons is the subject of many studies (e.g. that of chlorine organic molecules by Newton *et al.*, 1990; alkyl sulphates by Belanger *et al.*, 1995; pesticides by Allan *et al.*, 2004, 2005; mercury by Runck, 2007; copper by Serra *et al.*, 2009; and antifouling agents by Wahl *et al.*, 2010; Tlili *et al.*, 2011; Villeneuve *et al.*, 2011). EPS molecules are able to adsorb substances which are potentially toxic to biofilms such as Cd, Pb, Cu and Sr and this is due to their negative surface charge acquired from similarly charged functional groups which deprotonate as pH increases (Konhauser, 2007). Studies on cyanobacteria and Sulphate Reducing Bacteria (SRB) have revealed that the functional groups include carboxylic acids (R–COOH), hydroxyl groups (R–OH), amino groups (R–NH$_2$), sulphate (R–O–SO$_3$H), sulphonate (R–SO$_3$H), and sulphydryl groups (–SH), all of which bind metal ions including Ca^{2+} and Mg^{2+} (Dupraz *et al.*, 2009 and references therein).

This chelation ability can significantly alter the hydrochemistry of the surrounding water. For example, Li *et al.* (2010) and Ortega-Murales *et al.* (2006) found that cyanobacterial EPS bound 55 mg Ca^{2+} g^{-1} and 183 mg Ca^{2+} g^{-1}, respectively and that from SRB bound 120–150 mg Ca^{2+} g^{-1} (Braissant *et al.*, 2007). Removal of this amount of Ca^{2+} would be significant for flume experiments studying calcite precipitation.

Laboratory experiments assessing the impact of dissolved heavy metal concentrations on higher organisms need to take into account the ability of biofilms to accumulate these contaminants as bioaccumulation may significantly alter concentrations in the experimental solution. However, the impact of this is complicated by the substantial variations between organisms' capacities for a given metal. For example, biosorption of Cd was 8.0 mg g^{-1} for *Pseudomonas putida* (Pardo *et al.*, 2003) and 250.0 mg g^{-1} for *Staphyloccous xylosus* (Ziagova *et al.*, 2007), biosorption of Pb was 79.5 mg g^{-1} for *Pseudomonas aeruginosa* (Chang *et al.*, 1997) and 567.7 mg g^{-1} for *Corynebacterium glutamicum* (Choi & Yun, 2004) (see review of Wang & Chen, 2009 for further details). It is worth noting that the biosorption ability of biofilms is not limited to the living organisms. Inactive or dead components have been shown to accumulate metal ions through various physicochemical mechanisms (Wang & Chen, 2009). Moreover, sterile EPS extracted from biofilms also shows a strong ability to both sorb metal ions and alter the rate of mineral precipitation in the absence of cells (Saunders, 2012).

Other studies have used biofilms as potential functional indicators for the overall health of the hydro-ecosystem (e.g. the comparison by Claret & Fontvieille, 1997; impact of heavy metals by Dasilva *et al.*, 2009; and impact of pesticides by Rimet & Bouchez, 2011). Laboratory experiments which investigate ecosystem responses to changes in environmental factors such as hydrodynamics, nutrient levels and hydrochemistry also need to take into account the potential impacts of biofilms on higher level organisms. For example biofilms in which algae dominate over bacteria are unlikely to give adequate nutrition to detritivorous snails (Sheldon & Walker, 1997).

5.3.2   The impact of biofilms on the hydrodynamic environment

For fluvial biofilms growing on artificial or small-scale sediment beds, vertical velocity profiles and/or turbulence statistics may be sufficient to characterise interrelationships with local flow conditions. This approach was successfully used by Biggs & Thomsen (1995) to study the resistance of different biofilm communities to increasing shear stress, Nikora *et al.* (1997) to quantify the impact of biofilms on turbulent boundary layer structure and hydraulic roughness, and Hondzo & Wang (2002) to infer the relationship between biofilm growth dynamics and local flow conditions. In these studies, the most striking result was the ability of biofilms to modify the turbulent boundary layer, specifically the roughness height z_0 or the equivalent hydraulic roughness k_s. However, this approach can only provide insights about the near-bed local flow conditions if the bed is flat and homogeneous. Very few field studies (see review Biggs *et al.*, 1998) have attempted to characterise the flow close to the bed, but those studies that have, including the work of Biggs *et al.* (1998) that recorded velocity measurements to within 2 mm of the substrate using hot-film anemometers, clearly showed that near-bed velocities were far better correlated to the dynamics of biofilm than gross measures of velocity, such as the velocity at 0.4 of the depth (assumed to approximate the depth-averaged velocity; Biggs & Hickey, 1994; Takao *et al.*, 2008), cross-sectional average velocity or discharge per unit depth (Cazelles *et al.*, 1991; Kim *et al.*, 1992; Claret & Fontvieille, 1997; Biggs *et al.*, 1997; Ghosh & Gaur, 1998; Battin & Sengschmidtt, 1999; Robinson *et al.*, 2004; Lumborg *et al.*, 2006; Luce *et al.*, 2010a, 2010b) or total discharge (Matthaei *et al.*, 2003; Boulêtreau *et al.*, 2006).

Owing to the inherent heterogeneity and the intrinsic three-dimensionality of near-bed flow associated with the development of the roughness-sublayer above coarse, rough, substrates variably coated with biofilm, a methodological improvement was necessary to correctly describe near-bed flow. To this end, the pioneering work of Raupach (1991) for atmospheric flows over vegetal canopies using Double-Averaged Navier-Stokes equations (DANS), based on temporal and horizontal spatial averaging, was adapted by Nikora *et al.* (2002b) for river flows (see Nikora *et al.*, 2007a, 2007b and Chapter 4 in this publication for summaries). Nikora *et al.* (2002b) used ADV measurements and double-averaging to investigate the impact of biofilm on the structure of the turbulent boundary layer above cobble-like artificial substrates. They showed that both the displacement height z_d and the roughness height z_0 were modified by the biofilm, reducing turbulence levels relative to bare substrates. In recirculating flumes with low turbidity and larger-scale rough beds (cobble or gravel-like artificial substrates), Godillot *et al.* (2001) and Labiod *et al.* (2007) coupled vertical profiles of velocity obtained using an LDA with the double-averaging methodology to establish relationships between local velocities and biofilm growth dynamics. Graba *et al.* (2010) adopted a similar measurement protocol and used the results to significantly improve the performance of the growth model of Uehlinger *et al.* (1996) by replacing the cross-sectional average velocity with local flow quantities in the sloughing term. These studies also show how biofilms, in the absence of growth-limiting factors, gradually cause the hydraulic roughness of the bed to transition from being dominated by the shape of the underlying substrate to reflecting the morphology of the biofilm (porous mat and filamentous algae).

5.3.3 The impact of biofilms on the sedimentary environment

The stability of sediments is an important feature in ecosystems subjected to physico-chemical gradients such as those in intertidal or fluvial environments (Passarelli *et al.*, 2012). Several studies have investigated the interaction of biofilms with suspended matter. In cohesive and non-cohesive sediments, the EPS of established biofilms increases erosion thresholds by a factor of five and reduces detachment rates through the binding of sediment grains, changes boundary layer roughness and creates an adaptable structural matrix (Le Hir *et al.*, 2007; Grabowski *et al.*, 2011). Additionally, biofilms impart a protective skin to the sediment surface (Paterson *et al.*, 2000), with larger grain sizes requiring a thicker biofilm for protection (Le Hir *et al.*, 2007). It has been recognised for more than a decade that the exopolymers of biofilms are very efficient agents of biostabilization in cohesive sediments (see Uncles, 2002 for a review of estuarine research highlighting the impact of biofilms, deBrouwer *et al.*, 2000 for an illustration of the interplay between flow, biofilms and sedimentology, and Yallop *et al.*, 2000 for a field study of the relationship between biofilm biomass and biostabilisation). Numerous studies conducted in mesocosms and/or annular flumes have quantified the erodibility of sediment beds colonised by biofilms of different types and ages (see Sutherland *et al.*, 1998; Milburn & Krishnappan, 2003; Droppo *et al.*, 2007; Neumeier *et al.*, 2007; Droppo, 2009) and/or the nature of the eroded particles or coagulates (see Stone *et al.*, 2008).

The presence of biofilms on submerged sediments alters concentration gradients in the pore water, nutrient fluxes at the sediment – water interface and the hydro-chemistry of the overlying water (Woodruff *et al.*, 1999). A laboratory flume experiment by these authors revealed that biofilm development led to substantial changes in shallow (0–2.5 mm) sediment composition (Table 5.2) over an 8 week period.

Fine sediment loads are known to increase as a result of many types of land use change, damaging the riparian ecosystem (e.g. Wood & Armitage, 1997). Significant interest in the use of dams to manage flows and flush fine sediment through the fluvial system has resulted in numerous studies that consider the general ecological impacts of flushing, discussing biofilms as a component of the wider ecosystem (see Osmundson *et al.*, 2002; Collier, 2002; Jakob *et al.*, 2003; Robinson *et al.*, 2004; Dukowska *et al.*, 2007; Takao *et al.*, 2008; Fuller *et al.*, 2011). Other studies have focussed solely on biofilms and the roles of light attenuation (Davies-Colley *et al.*, 1992), silt aggradation (Graham, 1990) and abrasion by larger particles (Luce *et al.*, 2010a, 2010b) in altering biofilm dynamics. However, while it is possible to isolate the impact of biofilm removal on bed sediment erosion and transport in very energetic flows, it is difficult to separate the effects of the many external factors associated with suspended matter (e.g. light attenuation, clogging of pore spaces, abrasion). This has motivated the use of (usually recirculating flume-type) mesocosms to investigate both the impact of suspended sediments on biofilms (Francoeur & Biggs, 2006; Birkett *et al.*, 2007), and the effects of biofilms on deposition rates (Arnon *et al.*, 2010; Salant, 2011).

Luce *et al.* (2010a, 2010b) were able to use field data to show that saltation of sand particles may be an important driver of the abrasion of periphytic biofilms. However, the process-based model they propose (saltation and particle impact) could be significantly improved with laboratory flume observations and measurements. Also, the mechanical processes involved in sediment uptake are still not adequately quantified. The study of Salant (2011) is a very convincing example of a successful integrative approach, to the clogging and uptake of suspended sediment by river biofilms. Similar studies focusing on flow and sediment interactions and making use of the double-averaging methodology are still necessary to better understand the interplay between biofilms and sediments on both fine and coarse substrates.

Table 5.2 Comparison of the mineral composition of the top and middle sediment sections with the initial mixed sediments. The clay composition (illite, expandable and kaoloite) are expressed as the percentage composition of the <2 μm particle size fraction (Woodruff *et al.*, 1999).

Depth (mm)	Mica	Chlorite	Quartz	Vivianite	Calcite	Dolomite	Illite	Expandable	Kaolinite
0.5	–	3	60	15	21	1	36	50	14
1.5	–	2	81	6	10	1	33	49	18
2.5	–	4	84	3	8	1	38	46	16
17	2	3	71	20	4	–	33	52	15
Initial mixed	–	3	69	22	3	3	38	45	17

5.4 CONCLUSIONS

Biofilms are ubiquitous features of aquatic environments and have significant impacts on the hydrodynamics and sedimentary and chemical fluxes in marine and freshwater systems. Although the major external factors driving biofilm dynamics have long been identified and thoroughly quantified in field studies, there is a lot of scope for well-designed hydraulic experiments incorporating biofilms. Integrative approaches that simultaneously investigate biofilm composition, structure, function, dynamics, flow and sediment transport properties at the local scale, before extrapolating to the larger scale through numerical modelling, have emerged only recently and are mainly laboratory flume based. These very successful approaches have highlighted the limitations of field studies using only integrated flow descriptors, and the considerable advantages of using mesocosm studies in hydraulic laboratories to deploy highly sensitive measurement techniques. These have enabled the measurement of variables that are difficult to quantify in the field, such as solute or suspension removal and production rates (e.g., Gainswin et al., 2006a, 2006b; Arnon et al., 2007, 2010; Augspurger & Kusel, 2010), step injections of conservative tracers (Bottacin-Busolin et al., 2009), community succession from bare substrates and the analysis of biofilm structure (Besemer et al., 2007; Besemer et al., 2009b) and oxygen profiles (De Beer et al., 1996). There is growing recognition of the close relationship and positive feedback between modifications to boundary layer hydraulics (see Biggs & Thomsen, 1995; Nikora et al., 1997; Godillot et al, 2001; Nikora et al., 2002b; Hondzo & Wang, 2002; Labiod et al., 2007; Graba et al., 2010) and biofilm-mediated chemical precipitation of dissolved material (Hammer et al., 2008). Hydraulic experiments also have a role in resolving long-standing questions regarding the structure and significance of naturally-occurring mineral deposits (e.g. tufa, travertine) and the spatio-temporal patterns in constructed systems designed to remediate soluble pollutants (Mayes et al., 2009). However, care is needed if researchers are to successfully isolate the factors they wish to study. The difficulty of transferring intact biofilms from the field into the laboratory means that most studies grow the layers of organisms in situ. However, the impact of environmental factors on growth, form and character must be considered very carefully if the results of any laboratory experiments are to be meaningful.

Chapter 6

Plants, hydraulics and sediment dynamics

*M. Paul, R.E. Thomas, J.T. Dijkstra, W.E. Penning &
M.I. Vousdoukas*

6.1 INTRODUCTION

Most studies on coastal and fluvial hydro- and morpho-dynamics have focused on flow properties and their interactions with sediments. However, it has long been accepted that vegetation is also an important factor that can affect the flow in, and the morphology of, aquatic systems. Vegetation in shallow water systems, such as rivers and coasts, plays an important role in altering flow resistance and turbulence (Nepf & Vivoni, 2000; Ghisalberti & Nepf, 2002) and, consequently, it affects the transport of sediment, nutrients, and contaminants (Kemp *et al.*, 2000; Cundy *et al.*, 2005). Because these factors exert major influences on habitats and biodiversity (Leonard & Luther, 1995), aquatic and riparian vegetation has recently been the subject of important research into the management, preservation and restoration of freshwater and coastal ecosystems, including salt marshes (Deloffre *et al.*, 2007; French *et al.*, 2000) and floodplains (Baptist *et al.*, 2006). Although much of this work is field based there is a growing interest in using laboratory flume experiments to obtain detailed measurements of plant-flow interactions.

The primary mechanism by which flowing water exerts a force upon vegetation is through drag. Submerged or emergent vegetation reacts to the drag of water by either remaining erect, oscillating in response to turbulent fluctuations, or bending (Section 6.2.2). The magnitude of the drag force is a function of plant flexibility, frontal projected area, relative depth of submergence, and density (Li & Shen, 1973; Petryk & Bosmajian, 1975; Pasche & Rouvé, 1985; Fathi-Moghadam & Kouwen, 1997; Nepf, 1999; Freeman *et al.*, 2000; Stone & Shen, 2002; Järvelä, 2002, 2004; Wilson *et al.*, 2003, 2006b; White & Nepf, 2008), which may all vary with both plant type and age. All of these characteristics must be considered very carefully when designing laboratory experiments. According to Newton's Second Law, vegetation must also affect flow patterns. It does so by adding roughness and hence reducing velocity and attenuating turbulence in vegetated areas (Fonseca & Fisher, 1986; Fonseca & Koehl, 2006). Vegetation may thus increase flood risk (Green, 2005). However, it can also play an important role in preventing bank erosion in rivers, estuaries and along the shorelines of beaches and large lakes by i) deflecting flow away from banks (McBride *et al.*, 2007; White & Nepf, 2008; Hopkinson & Wynn, 2009) or dissipating wave energy (Möller *et al.*, 1999a) and ii) reinforcing the soil with roots and/or rhizomes (Coops *et al.*, 1996; Simon & Collison, 2002; Pollen & Simon, 2005; Thomas & Pollen-Bankhead, 2010). As a result, planting vegetation at the toe of stream banks and transplanting sea grass beds have been proposed as "soft" erosion protection measures (e.g. Price *et al.*, 1968; Allen & Leech, 1997; Bentrup & Hoag, 1998). Conversely, plants also introduce turbulence and induce scour along the interface between the vegetation and open-water (McBride *et al.*, 2007; Yang *et al.*, 2007; White & Nepf, 2008; Hopkinson & Wynn, 2009). Vegetation development, flow reduction and soil erosion are thus strongly interrelated processes in exposed riparian and coastal zones (Figure 6.1).

There are therefore many reasons for studying plant-flow interactions and it is not surprising that there is an extensive literature on the topic. However, given the variety of plants and their temporal and spatial variability, in addition to the wide range of flow and morphological conditions in nature, significant knowledge gaps still exist. Plants also impact on flows at a range of spatial and temporal scales, from boundary-layer processes and turbulent structures, through altering hydro- and morpho-dynamics around a plant stand, to altering river channel planforms at the largest spatial scales. This chapter presents advice to any researcher contemplating a laboratory experimental approach to exploring plant-flow interactions at the scale of interest. It reviews past work and incorporates suggestions as to what needs to be considered when designing ecohydraulic experiments that involve live plants.

The following terms defined here are used repeatedly throughout this chapter. A *stand* is a general term that is used to describe vegetated areas, including those associated with meadows and crown-building plant species. A *crown* refers to the aboveground part of a single plant that consists of a stem or stipe that supports leaves

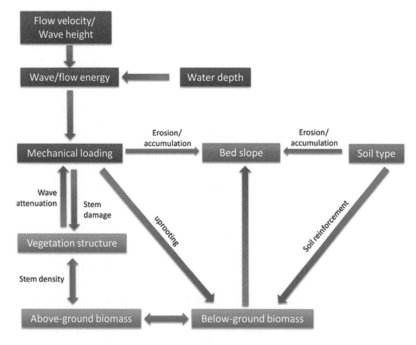

Figure 6.1 Schematic representation of the impact of flow/wave action on emergent plant growth and bed morphology (modified after Coops *et al.*, 1996).

or fronds at a certain distance above the bed. Crown-building species contain most of their biomass in the upper part of the canopy. A *meadow* describes an area that is covered by grass-like vegetation. Meadow-building species do not have stems or stipes, but leaves emerge from a sheath very close to the bed. Their aboveground biomass is evenly distributed along the vertical axis of the canopy. The *canopy* is the part of a plant stand that is aboveground and includes stems and stipes, as well as leaves and fronds.

6.2   THE IMPACTS OF PLANTS ON THE PHYSICAL ENVIRONMENT

6.2.1   Overview

Current literature shows that vegetation can increase flow resistance, control the mean and turbulent flow structure in many aquatic environments, and thereby modify sediment erosion, transport and deposition (e.g. Myrhaug *et al.*, 2009; Chen *et al.*, 2011; Feagin *et al.*, 2011), with implications for ecological processes, such as nutrient transport and pollen dispersal (Verduin *et al.*, 2002). Numerous studies have used field and laboratory experiments and/or numerical modelling to investigate specific flow conditions and habitats found in nature. In particular:

- River/open channel flows, where unidirectional flow is acting upon submerged/ emergent flexible vegetation.
- Propagation of extreme, large-scale, flow events (e.g. floodplain inundation, tsunamis and storm surges) over vegetation where the flow is unidirectional and acts upon flexible vegetation, but also on woody vegetation, such as floodplain forests and mangroves.
- Flow in the upper littoral zone with submerged/emergent flexible/rigid vegetation, under the influence of both waves and currents, related mostly to shorelines in shallow lakes, lagoons and estuarine environments.
- Wave-vegetation interactions in the sub-littoral zone, associated with kelp and sea grass mostly found on the seabed, or along the shoaling zone, and submerged flexible macrophytes in shallow freshwater systems.

Plant species that interact with water flows vary in morphotype, ranging from simple blade-shaped sea grasses to complex trees, with roots, branches and leaves. The plant attributes most important for flow studies are outlined below.

6.2.2   Classification of vegetation forms under the influence of flow

The behaviour of simple flexible aquatic vegetation (e.g. sea grass) forced by steady flow has been classified into the following three types (Ciraolo *et al.*, 2006; Yagci *et al.*, 2010):

i   *erect*, with the plants taking up a position which is steady and slightly inclined with respect to the vertical;

ii   *waving* (Figure 6.2), with the plants being considerably inclined and oscillating (*monami* or *honami*). Nepf & Vivoni (2000) segregated this class into two sub-classes, *gently swaying* and *strong coherent swaying*;

iii *prone* (Figure 6.2), with the plants taking up an inflected, sub-horizontal and quasi-steady position.

Conversely, up to six modes were identified for more complex plants with stems, branches and leaves as stream power increased (Gourlay, 1970; Pitlo & Dawson, 1990; Green, 2005):

i   at the lowest stream powers, vegetation is not deflected and is stationary;

ii   stems and leaves orientate themselves in the dominant flow direction;

iii stiff vertical stems vibrate and oblique or elongated horizontal stems begin to exhibit oscillatory movements (*monami* or *honami*);

iv   stiff stems become inclined and submerged leaves become strongly orientated, with surface leaves submerging or tipping and with a progressive loss of dead parts;

v   stems become prone or densely compacted, with surface leaves submerged;

vi   at the highest stream powers, damage and loss occurs of parts of, or entire plants.

However, the precise identification of the corresponding thresholds of stream power is problematic because of within-species, inter- and intra-location and seasonal variability. Likewise, the capacity of vegetation elements to withstand damage or to become uprooted is difficult to predict since it varies among individual plants, with season and with substrate conditions (Aranguiz *et al.*, 2011; Pollen-Bankhead *et al.*, 2011).

Under wave motion, flexible aquatic vegetation is known to sway in an oscillatory fashion, moving back and forth with the orbital water motion (Augustin *et al.*, 2009; Bradley & Houser, 2009). Depending on vegetation stiffness and wave forcing this motion can be whip-like or comparable to a cantilever (Denny *et al.*, 1998; Denny & Gaylord, 2002). The occurrence of the above modes is known to be controlled by the magnitude of the forcing (i.e. the drag imparted by the flow) and the biomechanical characteristics of the plant. These will be discussed in the next two sections. The physical forcing of wave action and water flow leads to changes in the structure and

 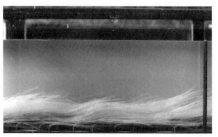

Figure 6.2 Example of waving (left) and prone (right) vegetation (*Posidonia oceania*) under unidirectional flow (Ciraolo *et al.*, 2006).

composition of plant stands over time, resulting in an unsteady interaction of fluid flow with aquatic vegetation (Méndez & Losada, 2004).

6.2.3   Quantifying the drag force acting on vegetation

To account for the enormous spatial variability observed in turbulent flows inside and above atmospheric canopies (i.e. in the roughness sub-layer), Wilson & Shaw (1977) and Raupach & Shaw (1982) introduced the concept of spatial averaging the equations of turbulent fluid motion. This formulation, since adapted by Nikora *et al.* (2001) for use in hydraulics, is a rigorous formulation of the time- and space-averaged flow in the roughness sub-layer with explicitly derived terms for drag and dispersion. This "double-averaged" concept is now increasingly applied in the aquatic context, as exemplified in the review of Nepf (2012). The double-averaged momentum equations for turbulent flow can be written as (Nikora *et al.*, 2007a):

$$\frac{\partial \phi \langle \overline{u_i} \rangle}{\partial t} + \phi \langle \overline{u_j} \rangle \frac{\partial \langle \overline{u_i} \rangle}{\partial x_j} = \phi g_i - \frac{1}{\rho} \frac{\partial \phi \langle \overline{P} \rangle}{\partial x_i} + \frac{\partial}{\partial x_j} \phi \left(\mu \left\langle \frac{\partial \overline{u_i}}{\partial x_j} \right\rangle - \langle \overline{u_i' u_j'} \rangle - \langle \overline{u_i'' u_j''} \rangle \right)$$

$$- \frac{1}{V_0} \iint_S \left(\mu \frac{\partial \overline{u_i}}{\partial x_j} \right) n_j dA_0 + \frac{1}{\rho_w} \frac{1}{V_0} \iint_S \overline{P} n_i dA_0 \qquad (6.1)$$

where ∂ is the partial differential operator, d is the differential operator, ϕ is the porosity which has a specific definition given by Nikora *et al.* (2007a), $\langle \overline{u_i} \rangle$ is the time- and space-averaged velocity, t is the time, g is the acceleration due to gravity (≈ 9.81 m s^{-2}), ρ_w is the mass density of water (≈ 1000 kg m^{-3}), P is the pressure, μ is the dynamic viscosity of water ($\approx 1.4 \times 10^{-3}$ N s m^{-2}), V_0 is the total volume of the averaging domain, $\mathbf{n}$ is the outward unit vector normal to the water-bed interface, A_0 is the area of the water-bed interface bounded by the averaging domain, the subscripts i and j range from 1 to 3, corresponding to two out of three orthogonal velocity components u, v, and w in the three coordinate directions, x, y, and z, and overbars represent time-averaged values, primes refer to fluctuations about these values, while angled brackets denote spatial averages and double primes represent fluctuations about those averages.

The turbulent stresses, $-\langle \overline{u_i' u_j'} \rangle$, can be parameterised by the multi-equation aniso-tropic model (Naot *et al.*, 1996; Choi & Kang, 2004), the two-equation isotropic model (Shimizu & Tsujimoto, 1994; Lopez & Garcia, 1998; Neary, 2003; Leu *et al.*, 2008) or the one-equation model (Li & Yan, 2007). The double-averaged equations contain additional terms compared to the equivalent time- (or Reynolds-) averaged equations: dispersive or form-induced stresses, $-\langle \overline{u_i'' u_j''} \rangle$, due to spatial fluctuations, and the total drag force due to plants and/or roughness elements that is composed of both viscous drag:

$$- \frac{1}{V_0} \iint_S \left(\nu \frac{\partial \overline{u_i}}{\partial x_j} \right) n_j dA_0, \qquad (6.2)$$

and form (pressure) drag:

$$\frac{1}{\rho}\frac{1}{V_0}\iint_S \overline{Pn_i dA_0} \tag{6.3}$$

At present, equations have not been developed to parameterise form-induced stresses, but the drag force terms may be parameterised using the Darcy–Forchheimer equation for flow through porous media (Dullien, 1979; Nield & Bejan, 1999):

$$F_D = -\left(\frac{v\phi}{K}\langle\overline{u_i}\rangle + \left[\frac{Y\phi}{K^{1/2}}\right]\langle\overline{u_i}\rangle\big|\langle\overline{u_i}\rangle\big|\right) \tag{6.4}$$

where F_D is the drag force, K is the permeability of the porous medium and Y is a non-linear momentum loss coefficient or inertial factor. For a given fluid, at very slow velocities, the first term on the right hand side of Equation 6.4, representing skin friction, becomes dominant and this equation reduces to Darcy's law:

$$F_D = -\frac{v\phi}{K}\langle\overline{u_i}\rangle \tag{6.5}$$

However, at faster velocities, the second term on the right hand side of Equation 6.4, representing form drag, becomes dominant and Equation 6.4 reduces to:

$$F_D = -\left[\frac{Y\phi}{K^{1/2}}\right]\langle\overline{u_i}\rangle\big|\langle\overline{u_i}\rangle\big| \tag{6.6}$$

Equation 6.6 can also be written in the more familiar form of the classical drag equation:

$$F_D = -\frac{1}{2}C_D A_p \langle\overline{u_i}\rangle\big|\langle\overline{u_i}\rangle\big| \tag{6.7}$$

where C_D is the isotropic drag coefficient and A_p is the frontal (projected) area of the plant per unit volume.

Significant simplification can be made in unidirectional flows, and Dunn et al. (1996) were able to obtain a backwater curve for open-channel flow through emergent vegetation. They found that the mean drag coefficient $\overline{C_D}$, for patches of vegetation could be estimated using:

$$\overline{C_D} = 2gh\frac{S_0 - S_f - \dfrac{dh}{dx}\left(1 - \beta\dfrac{[Q/A]^2}{gh}\right)}{A_p h\beta[Q/A]^2} \tag{6.8}$$

where S_f is the friction slope estimated using a uniform flow equation, Q is the flow discharge, A is the flow area and β is a coefficient accounting for the vertical distribution of streamwise velocity ($\approx 1 + ff/8\kappa^2$ if the von Kàrmàn-Prandtl law of the wall holds throughout the flow depth; Lin & Falconer, 1997), ff is the Darcy-Weisbach roughness coefficient, and κ = von Kàrmàn constant (≈ 0.33 in suspended sediment-laden flows; Bennett *et al.*, 1998).

It is stressed that despite obvious similarities and successful technology transfer between the two fields, two fundamental differences exist between canopies in the atmospheric and aquatic environment: i) aquatic canopies often occupy a larger proportion of the boundary layer than atmospheric canopies (Nepf, 2012), and ii) aquatic vegetation can be positively buoyant, which may provide a significant contribution to the resistance to hydrodynamic drag in addition to stiffness which occurs in both environments (Luhar & Nepf, 2011). However, the balance between and impact of both contributions is still not well understood and merits further investigation.

In bi-directional flows, C_D has commonly been estimated as a function of the Reynolds number (Kobayashi *et al.*, 1993; Mendez *et al.*, 1999):

$$\overline{C_D} = \omega + \left(\frac{\xi}{Re_\nu}\right)^{\psi^3} \tag{6.9}$$

where ω, ξ and ψ are coefficients and Re_ν is the stem Reynolds number based on the plant diameter or blade thickness d and a characteristic velocity acting on the plant u, and defined as the maximum value at the top of the vegetation. At high Reynolds numbers the drag coefficient becomes asymptotic (Figure 6.3).

Mendez & Losada (2004) found a stronger relationship between C_D and the Keulegan–Carpenter number KC, defined as KC = uT_p/d, where u is a characteristic velocity acting on the plant, defined as the maximum horizontal velocity in the middle of the vegetation stand and T_p is the peak wave period. However, for flexible sea grasses, Bradley & Houser (2009) found the strongest correlation between C_D and Re_ν. A recent laboratory study using rigid stems (Augustin *et al.*, 2009) showed that the C_D-Re_ν relationship was strongest when stems were emergent but that the C_D-KC relationship was strongest when stems were submerged

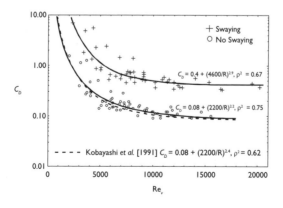

Figure 6.3 Drag coefficient C_D against the stem Reynolds number Re_ν. Comparison of field measurements with different empirical coefficients (Mendez *et al.*, 1999).

or near-emergent. Myrhaug *et al.* (2009) used results from these empirical studies to develop an analytical method to estimate the drag force on a stand of vegetation subjected to nonlinear waves (through Stokes' second order wave theory). However, their results have not yet been tested against either field or laboratory measurements.

6.3   IMPORTANT PLANT CHARACTERISTICS FOR PLANT-FLOW INTERACTIONS

Several factors have been shown to affect the behaviour of a plant in a flow field (Green, 2005) and all of these must be considered carefully when designing laboratory experiments. They can be grouped into properties of the individual plants and of the plant stand. At the individual plant level the following factors are important:

- Plant and leaf dimensions, shape and morphotype;
- Stiffness and elasticity;
- Buoyancy.

At the plant stand level:

- Density and distribution of the plants;
- Seasonal occurrence/variations of the plants;
- Different types of plants;
- Spatial occurrence of plants parallel/normally to the direction of flow (e.g. at the centre or close to the channel bank/at pools or elevated sections);
- Maximum level of growth that each species can reach in a cross-section and with respect to water depth.

6.3.1   Plant properties

Neumeier (2007) found that the roughness length of vegetation is not related statistically to flow velocity nor water depth, but depends only on the characteristics of the canopy. This conclusion is different from his previous studies that showed dependence either on the flow velocity (Neumeier & Amos, 2006) or on the vertical biomass distribution (Neumeier & Ciavola, 2004). However, it shows clearly that properties like the plant form and stiffness (as well as density) are critical for the hydraulic resistance of individual plants or vegetation stands.

6.3.1.1   Plant shape

Plant form has been underlined as an important factor for the effect of vegetation on flows (Li & Shen, 1973; Petryk & Bosmajian, 1975; Haslam, 1978). While a great variety of plant shapes exist, two general groups can be identified: meadow building species where the biomass is evenly distributed along the vertical axis of the plant (e.g. most sea grass species) and crown building species with a stipe or stem and the majority of biomass in the upper part of the plant (e.g. trees).

Under wave motion, both plant shapes have been observed to reduce wave energy and shear stress near the bed (James & Barko, 2000). However, their presence leads to different velocity profiles with meadow building plants yielding a steady decrease of orbital velocity with depth (Koch & Gust, 1999) and crown building species producing a more complex profile with maximum energy values just above the top of the canopy (Verduin & Backhaus, 2000).

Under unidirectional flow, flexible plants with stipes have been shown to result in 50% greater velocity reduction compared to simple flexible rods with or without stipes (Wilson et al., 2003). Comparisons of both plant forms showed that the additional superficial area of the fronds alters the momentum transfer between the within-canopy and surface flow regions by a) inducing larger drag forces; and b) reducing shear-generated turbulence due to the inhibition of momentum exchange by the frond surface area (Wilson et al., 2003).

6.3.1.2   Stiffness

Plant stiffness is critical for whether plants will tend to compress, bend or oscillate under the effect of a drag force. The behaviour of stiff plants can be considered similar to that of rigid vertical cylinders or can be described by a cantilever motion. Plants that flex easily, on the other hand, bend under hydrodynamic forcing and may move in a whip-like fashion under waves.

In unidirectional flow, plants lead to a reduction of the mean velocity and a change in the velocity profile. It has been shown that the effect of stiff plants is more pronounced as they require a higher forcing and hence flow velocity to bend. Moreover, the effect of stiff vegetation is more dependent on vegetation density compared to highly flexible sea grass shoots (Fonseca et al., 1982; Gambi et al., 1990). The main reason is turbulence, since vortices above a still canopy have been shown to rotate faster than those above a waving canopy, with the latter being more effective in turbulence attenuation (Chen et al., 2011).

Studies on large macroalgae (Koehl, 1996; Denny & Gaylord, 2002), salt marsh (Bouma et al., 2010) and sea grass (Paul et al., 2012) have shown that the wave attenuating capacity increases with increasing stiffness. But, as is the case under unidirectional flow, a dependence on shoot density was observed. Flexible plants can dissipate wave energy at higher shoot densities as effectively as stiff plants at low densities (Bouma et al., 2010; Paul et al., 2012) and Bouma et al. (2010) suggested a dependence on biomass rather than on stiffness or density alone.

The effect of stiffness also depends on the hydrodynamic forcing. Under unidirectional flow, plants will bend in order to streamline, but a stiffer plant will require higher flow velocities to bend to the same extent as more flexible plants. Very stiff vegetation may even break or be uprooted before bending can be achieved. Under wave forcing, two types of motion need to be distinguished: cantilever and whip-like. Relatively flexible vegetation (e.g. sea grass) acts as a cantilever under low wave forcing but this changes to whip-like motion with increasing wave amplitude and/or period (Manca, 2010). The point of transition from one type of motion to the other depends on the stiffness of the plant. It is not yet known how the type of motion affects wave attenuation and whether it needs to be accounted for when assessing the stiffness of a plant (Stewart, 2006; Paul, 2011).

6.3.1.3 Buoyancy

Aquatic vegetation can be positively buoyant as a result of either having lacunae or pneumatocysts, a distinct difference from terrestrial plants. Buoyancy is generally higher in less stiff plants (Méndez & Losada, 2004) and like stiffness, buoyancy provides resistance against hydrodynamic drag. The effect of buoyancy on wave forcing is not yet understood, but for unidirectional flow Luhar & Nepf (2011) showed that the impact of buoyancy depends on the flow velocity and the location along the plant leaf. Studies with structurally simple sea grass leaves showed that bending and leaf posture close to the bed were independent of buoyancy while towards the leaf tips buoyancy added to the restoring force. However, with increasing flow velocities drag force exceeded leaf buoyancy and stiffness became the dominant restoring mechanism (Luhar & Nepf, 2011).

6.3.2 Vegetation stand properties

6.3.2.1 Areal stem density

Higher areal stem density is expected to result in additional momentum extraction via the hydrodynamic form drag of the vegetative "roughness elements" (e.g. Kadlec, 1990; Shi *et al.*, 1995). The laboratory experiments of Graham & Manning (2007) have shown that sequential increases in stem density cause an approximately exponential attenuation of mean current velocity. Other studies on salt marsh (Leonard & Croft, 2006; Neumeier, 2007) and sea grass (Peterson *et al.*, 2004; Widdows *et al.*, 2008a) confirmed a reduction of flow velocity with increasing shoot density.

A positive correlation between energy dissipation and vegetation density has also been observed for oscillatory motion under waves (Nepf *et al.*, 1997; Koch & Gust, 1999; Bouma *et al.*, 2005; Möller, 2006; Augustin *et al.*, 2009). However, a minimum shoot density was required before wave attenuation could be observed within the relatively small sea grass species *Ruppia maritima* (Newell & Koch, 2004) and *Zostera noltii* (Paul, 2011).

Shoot density of most (freshwater and saltwater) macrophyte species changes seasonally and the above results suggest that the effect of plant stands on flow conditions will therefore vary over an annual cycle (Morin *et al.*, 2000; Pollen-Bankhead *et al.*, 2011). This is particularly the case for floodplain areas where vegetation might develop over a period of a few years from bare soil and low grasses, to shrubs and forests (Asselman *et al.*, 2002; Baptist *et al.*, 2004). Also, the presence of leaves on woody floodplain vegetation, such as willows can increase the friction factor two- to threefold (Järvelä, 2002). Augustin *et al.* (2009) observed that the influence of shoot density on wave attenuation increased with decreasing submergence ratio (defined as the ratio of water depth to vegetation height) for artificial salt marsh. This was recently confirmed for *Posidonia oceanica* surrogates (Prinos *et al.*, 2010) and suggests that, along the coast, the impact of vegetation on wave energy changes over the tidal cycle.

6.3.2.2 Configuration

Apart from the vegetation density, plant configuration patterns have also been shown to be important for canopy turbulence and have been extensively investigated in the past (Finnigan, 2000; Poggi *et al.*, 2004; Nezu & Sanjou, 2008; Righetti, 2008; Pietri *et al.*, 2009; Chen *et al.*, 2011). Early studies of interaction between flow and vertical cylinders have shown that staggering the cylinders was much more effective in reducing flow rates than if the cylinders were directly aligned (e.g. Li & Shen, 1973). This is because when the cylinders are placed in rows, the retardation effects are restricted to those distinct bands, but the flow in the spaces between the rows is relatively unhindered. On the other hand, when the cylinders are arranged in a staggered pattern, retardation is more evenly distributed, thus preventing any part of the flow from accelerating (Figure 6.4). The wakes behind many aquatic macrophytes are also significant; Machata-Wenninger & Janauer (1991) recorded reduced velocities up to at least 6 m downstream of a 3 m wide stand of *Groenlandia densa* (opposite-leafed pondweed). Nezu & Sanjou (2008) found that larger streamwise and spanwise spacings between closely grouped vegetation elements resulted in reduced variability in spanwise velocity distributions and reduced the mean submergence depth. Chen *et al.* (2011) conducted extensive tests with different vegetation configurations and concluded that measured velocity profiles were significantly different between a) centred and staggered arrangements, and b) protected (behind stems) and exposed (between stems) locations.

6.3.2.3 Spatial distribution

Another important aspect is the ratio between the width of the vegetated area and the width of the flow, since this has been shown to significantly influence within

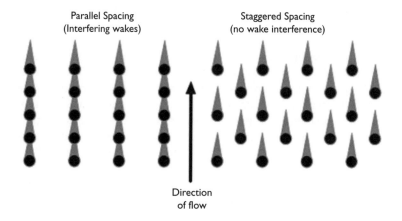

Figure 6.4 Example of plant configurations (Green, 2005). Black circles show stem locations.

and over-canopy flow behaviour, as well as within canopy turbulence intensity (e.g. Fonseca & Koehl, 2006). In nature vegetation patches are generally significantly narrower than the width of the flow domain, allowing extensive intermediate space for the diverged flow. However, there are cases when the flow is forced through the vegetation field and, especially under (near) emergent vegetation conditions, hydraulic resistance may be significantly increased (Bulthuis et al., 1984; Powell & Schaffner, 1991; Nepf, 1999; Serra et al., 2004), causing blockage or flooding (e.g. Haslam, 1978; Green, 2005) or even hydraulic jumps (Fonseca et al., 1982). Nowell & Jumars (1987) therefore concluded that the spatial distribution of vegetation patches in flume studies needs to match field conditions, i.e. if in the field the flow is expected to diverge around a patch, it should not be forced through the vegetation stand in the flume. Fonseca & Koehl (2006) suggested that the percentage of the area of an experimental facility occupied by vegetation must compare with natural conditions. Therefore, installing plants across the entire width of a flume is appropriate if stands are of a wide lateral extent relative to the incoming flow so that limited lateral deflection of flow around the patch or intrusion of flow from the edges of the stand are likely. Conversely, allowing flow to pass around as well as over a canopy in a flume may better simulate small patches (Fonseca & Koehl, 2006). Within channelised flows, changes to the lateral position of vegetation may significantly alter flow and sediment transport pathways (e.g. Tsujimoto, 1999; McBride et al., 2007; White & Nepf, 2008; Larsen et al., 2009; Pollen-Bankhead et al., 2011; Bennett et al., 2008; Zong & Nepf, 2010; Gurnell et al., 2012). In all cases, care must be taken during experiments to avoid the compression of flow paths and associated acceleration near the sidewalls of the flume.

6.3.2.4   Submergence ratio

The effectiveness of vegetation in attenuating waves has been shown to increase with the percentage of the water column it occupies (Ward et al., 1984; Fonseca & Cahalan, 1992; Koch, 2001; Koch et al., 2006; Möller, 2006). Therefore, the ratio of water depth to vegetation height, or submergence ratio, is an important control on flow reduction. In stiff vegetation emergent conditions are also possible and these are even more effective in attenuating waves (Knutson et al., 1982; Kobayashi et al., 1993; Lövstedt & Larson, 2010).

For submerged vegetation, Chen et al. (2007) showed that the upper depth limit for wave attenuation by vegetation depends on the incident hydrodynamic forcing. Their model suggests that wave period and orbital velocity determine how effectively sea grass attenuates waves in water depths too deep for the orbital motion of waves to reach the bed, but shallow enough for them to interact with the sea grass canopy. Recent studies suggest, however, that at least in shallow water conditions shoot density and leaf area index can compensate for a high submergence ratio (Bouma et al., 2010; Paul et al., 2012).

For homogeneous vegetation (e.g. sea grass, salt marsh) and unidirectional flow, the emphasis has been on flow velocities rather than on submergence ratio (Fonseca et al., 1982). In those cases, a layered velocity profile will develop. For vegetation that holds more biomass in the upper part of the plant (e.g. trees) submergence ratio can be important, as vegetation density increases substantially once the crown is submerged. Consequently, studies addressing the damping performance of mangroves on tsunamis applied different submerged volume ratios to each part of the mangrove forest: roots, trunk and canopy (Mazda et al., 1997; Harada et al., 2000).

6.3.3   Measuring plant properties

Traditional measures of plants used by biologists, including shoot density, ground coverage and canopy height, are insufficient to quantify the capacity of vegetation to interact with the fluid and sediments (Neumeier, 2005). The properties of plants with a simple morphology (e.g. sea grass) are easier to quantify, i.e. through the leaf length and the Leaf Area Index (LAI) which is the ratio of the leaf area to the ground area (Gacia & Duarte, 2001). For tall emergent plants like *Juncus roemerianus*, it is sufficient to measure the stem density and stem diameters, because the water essentially flows through vertical cylinders (Leonard & Luther, 1995). However, the canopy is often much more complex, either because the plant is composed of several parts with different shapes (e.g. stems, leaves, flowers), or because the canopy is composed of different species, individuals of different ages or even a mix of living and dead individuals.

Standard methodologies involve careful plant harvesting and laboratory analysis and can also include weighing to obtain the "above ground biomass". Vertical biomass distributions are typically obtained by harvesting a known "sampling area", counting the shoot/stem density (Neumeier, 2005) and measuring mean diameter and height (Feagin et al., 2011). Vertical layers are typically defined and weighed separately (after being cut) to give biomass profiles. Depending on the study, root systems can be harvested and weighed and the volume measured by displacement.

While the approach of harvesting and weighing plants in horizontal layers is a first step to take into account morphological complexity, it does not provide a direct relationship between the weight of a plant and flow resistance and it does not describe the canopy density at different heights, which is a key parameter for the flow structure. As a result, optical methods that have been successfully applied to heathland and trees have been used to provide additional information about vertical biomass distributions (Baker et al., 1996), floodplains (Straatsma et al., 2008) and terrestrial grassland (Zehm et al., 2003). The general concept is that lateral obstruction can be used as a proxy for the spatial distribution of vegetation (Möller, 2006; Neumeier, 2005) and towards that purpose, lateral pictures of a known volume of canopy are obtained against a coloured background. Image analysis and edge detection techniques are then applied to differentiate between vegetation and background. There are variations in the implementation of the approach, depending on the characteristics of the study area (e.g. background colour, volume of considered area, point and angle of view).

Despite several shortcomings, optical methods are increasingly being used, since they have key advantages over more traditional vegetation survey methods (Möller, 2006):

- They minimise time in the field (can be deployed quickly at pre-defined sites);
- They are non-destructive (although areas around the photographed plots are temporarily disturbed);

- They allow detailed, objective, and quantitative measurements of a series of vegetation structure parameters (e.g. canopy "roughness" and density) without measurement errors frequently encountered in direct field measurements;
- They are insensitive to operator error, optical distortions and technical problems.

One of the main difficulties in the application of the optical method is that the relation between percentage image obscuration and dry biomass is not linear once plant density exceeds a certain value. This is related to the fact that leaves deeper in the depth of view can be "sheltered" (shadowed) by those closer to the camera. For dense plants and stands this becomes an important issue as a further increase in biomass does not lead to a corresponding increase in percentage image obscuration (Neumeier, 2005; Möller, 2006). More recently, optical methods have been combined with terrestrial laser scanning (Antonarakis et al., 2009; Straatsma et al., 2008) which has the same problems with shadowing, but does allow accurate estimations of the position and shape of the vegetation.

The mechanical properties of plants can be expressed through Young's modulus of elasticity and the failure stress (bending strength). This involves conducting beam loading tests, preferably in the field and/or immediately upon harvesting of the stem sample (Feagin et al., 2011). Stems are usually fixed horizontally to a beam loading apparatus and an incrementally increasing load is applied to a single location on the stem. Stem deflection as a function of load is recorded, until stem failure (breakage) occurs (Fonseca & Koehl, 2006). Young's bending modulus can then be computed by:

$$E = \frac{J}{I} = \frac{Fa^2}{2\delta I(3L_s - a)} \tag{6.10}$$

where E is Young's modulus of elasticity (N m^{-2}), J is the flexural rigidity (N m^2), I is the second moment of inertia (for a cylinder, $I = \pi d_s^4/64$, in m^4), d_s is stem diameter (m), a is the distance from the base of the stem to the point at which F is applied (m), δ is the deflection of the stem (m), and L_s is the stem length. For example, Fonseca & Koehl (2006) measured E values around 4.7×10^{10} N m^{-2} for a plastic ribbon (surrogate), whereas live blades and sheaths of Zostera marina plants had E values of 8×10^8 and 1.7×10^7 N m^{-2}, respectively. One practical issue is that plants have heterogeneous, anisotropic, tapering stems. In addition, stems that have been loaded by unidirectional flows tend to be preferentially stiffer and/or more or less brittle in one direction. Multiple tests are therefore usually made in multiple directions and for multiple loading points.

Buoyancy may be measured using standard techniques that are applicable to other materials, although practical problems may arise for plants with complex structures. The force needed to sink a buoyant part of a plant can be measured by attaching strands with small weights (like floaters/bobbers on a rod), increasing the weight in steps and observing when the plant part sinks. However, different plant parts may have different buoyancy, i.e. the stems can be highly buoyant and the leaves of the same plant can be neutrally buoyant or even sink. Other plants (e.g. macro-algae) form dedicated air-chambers in their stems as floaters. Deriving a characteristic buoy-ancy value for a complex plant is therefore a challenging task.

6.4 THE IMPACTS OF PLANTS ON UNIDIRECTIONAL FLOWS

Unidirectional flow and turbulence affect plants through the transport of sediment, nutrients, seeds and toxins and impose a drag force on vegetation, which results in changes in plant posture as outlined above. Plants can adapt to these hydrodynamic conditions (e.g. Puijalon *et al.*, 2005), using various strategies to mitigate against the trade-off between drag avoidance, tolerance and access to nutrients. In unidirectional flow, the mean drag on plants is determined by the flow velocity close to and inside the canopy, with turbulent bursts causing a fluctuating component (Plew *et al.*, 2005). The temporal mean flow is also responsible for the advection of substances towards or away from the canopy, whereas turbulence largely determines the exchange of these substances between the bed, the plant tissue, the canopy and the mean flow.

Conversely, plants affect the flow by enhancing form drag and skin friction and by forming an obstacle and deflecting flow. Moreover, plant leaf litter can have an important impact on the chemical and physical environment of streams by provid-ing a source of carbon for stream biota and therefore fostering stream ecosystems (Ferreira *et al.*, 2010). While the importance of riparian leaf litter for stream function has been recognised (Encalada *et al.*, 2010), its impact on the physical environment has not yet been investigated and requires further study.

6.4.1 Impacts on the velocity profile

The presence of vegetation affects the velocity profile, the shape of which is a func-tion of flow velocity, plant properties and whether the vegetation is submerged or emergent. The velocity profile in the presence of submerged vegetation has been gen-erally divided into either two (Shimizu & Tsujimoto, 1994; Righetti & Armanini, 2002; Neary, 2003; Rowinski & Kubrak, 2003; Wilson *et al.*, 2006a) or three (Huai *et al.*, 2009) layers. In meadow-building species, an S-shaped profile has been identi-fied (Gambi *et al.*, 1990; López & García, 1998; Ghisalberti & Nepf, 2002; Ciraolo *et al.*, 2006; Neumeier & Amos, 2006), consisting of the following three flow zones (Figure 6.5):

Zone I, found inside the lower section of the vegetated layer, is characterised by low values of both velocity u and of its gradient $\partial u / \partial z$, mostly due to increased hydraulic resistance and turbulence generation (Fonseca & Koehl, 2006) from the denser vegetated field. Nepf & Vivoni (2000) have characterised this as the "lon-gitudinal exchange zone" as communication with the surrounding water is mostly through longitudinal advection.

Zone II, which can be of low thickness, is characterised by high velocity gra-dients, with the velocity increasing to almost a maximum value. This zone begins inside the vegetation, ends above its top, and includes a profile inflection point, found close to the vegetation top. Several studies have shown that the regions of highest Reynolds stresses exist in the upper part of the canopy (Leonard & Luther, 1995;

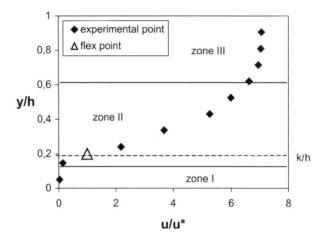

Figure 6.5 Typical velocity profile with three flow zones (Ciraolo *et al.*, 2006). The dashed line (k/h) represents the top of the vegetation canopy and the solid lines indicate the transitions between flows. The flex point highlights the position where the profile changes inflection. It is located close to the top of the canopy.

Leonard & Reed, 2002; Neumeier & Amos, 2006, Graham & Manning, 2007; Neumeier, 2007). The high velocity gradient is a consequence of the large difference in the velocities of Zones I and III. The flow characteristics have been shown to be analogous to those of a free mixing layer (Poggi *et al.*, 2004; Nezu & Sanjou, 2008); i.e. a confined layer that forms between two constant velocity layers, in which the shear does not arise from boundary conditions but from the difference in the velocities of the two constant velocity layers (Ghisalberti & Nepf, 2002; Ciraolo *et al.*, 2006). The velocity profile of Zone II closely follows a hyperbolic tangent-type distribution, whose parameters depend on the velocities in the other two zones (Ghisalberti & Nepf, 2002). However, other studies have also verified the applicability of the semi-logarithmic law in the upper part of Zone II, above the inflection point (Ciraolo *et al.*, 2006).

Zone III, which extends above the vegetation up to the free surface and is characterised by higher velocities, conveying a large percentage of the total flow. Nepf & Vivoni (2000) refer to Zones II and III as a "vertical exchange zone" where vertical turbulent exchange is very important for the momentum balance and turbulence. Depending on the total discharge, Turbulent Kinetic Energy (TKE) and shear can be high, while the shape of the velocity profile has been shown to follow a typical logarithmic (Klopstra *et al.*, 1997; Stephan & Gutknecht, 2002; Järvelä, 2005; Neumeier & Amos, 2006; Baptist *et al.*, 2007; Nezu & Sanjou, 2008) or power law (Cheng, 2007) influenced by the free water surface. Ciraolo *et al.* (2006) proposed an extension of the wake function of Finley *et al.* (1966); while other studies have characterised bent vegetation as a "secondary bottom" (e.g. Huai *et al.*, 2009) or as a form of macro-roughness (Neumeier & Ciavola, 2004).

The bending of plants under unidirectional flow can also lead to the onset of skimming flow with low velocity gradients (Neumeier & Amos, 2006). Under these conditions the flow glides over the closed plant canopy with less frictional coupling, leading to highly reduced mixing between the water above and within the canopy (Thompson et al., 2004). When the vegetation is emergent its entire area acts as "longitudinal exchange zone" (Nepf & Vivoni, 2000) and only Zone I exists (Neumeier, 2007).

It is generally accepted that increasing vegetation density causes greater flow retardation inside the canopy (Peterson et al., 2004; Leonard & Croft, 2006). However, several studies (e.g. Fonseca et al., 1982; Gambi et al., 1990) found no influence of shoot density on flow structure within the canopy. This conclusion is perplexing given the dependence of the drag force on projected area (Equation 6.7), implying significant overlap or sheltering and/or an inability of plants to fully adjust their form because of experimental limitations on flow area (i.e. flume dimensions). Nonetheless, this ambiguity also suggests the (presently unidentified) existence of parameter ranges in which vegetation density or potentially other vegetation parameters are irrelevant.

6.4.2 Impacts on turbulence generation and dissipation

Most experimental studies of turbulence near vegetation have focused on surrogates rather than real plants. Churchill (1988) showed that for a single erect cylinder in a horizontal flow field, flow is laminar for $Re_v < 150$ and for $6 < Re_v < 44$ separation occurs which creates a recirculation zone in the lee of the cylinder. For $44 < Re_v < 150$, organised vortex shedding is observed, while for $150 < Re_v < 30000$, a turbulent wake forms. In a multi-cylinder array, the limits of these regimes are likely to be different because of wake interference, sheltering and tortuosity (Green, 2005; Chen et al., 2011). For example, Nepf (1999) suggests that laminar flow in a multi-cylinder array occurs for $Re_v < 200$. Lateral shear produced by flow divergence around stems may delay the onset of vortex shedding and the consequent development of turbulent wakes in the lee of downstream stems.

Two important sources of turbulence can thus be identified in vegetation stands in unidirectional flow: i) wakes created by the stems, with a length scale approximately equal to the stem diameters (e.g. Li & Zhang, 2010) and ii) the shear layer above the vegetation, with larger eddies that can penetrate into the canopy (Shimizu & Tsujimoto, 1994; López & García, 2001). The bed boundary layer is a third source, but it is only relevant in canopies with very little cover near the bed. Dissipation of turbulence via the energy cascade occurs both inside and outside stands of vegetation; outside, dissipation is through viscous stresses only, while inside dissipation is accelerated by the stems that break down large-scale turbulence. Hence, both the generation and dissipation of turbulence depend on the vegetation size, shape and density.

Stands of plants introduce significant local velocity variation and the combination of this variability and turbulent diffusion leads to the process called mechanical dispersion (Nepf, 1999; Serra et al., 2004). Nezu & Onitsuka (2001) demonstrated that horizontal vortices near a free surface are caused by the inflection of shear instability and that turbulence near the free surface is transported laterally from the non-vegetated zone towards the vegetated zone by secondary currents. Maximum turbulence intensity and Reynolds stress occur at approximately the maximum deflected plant height

(Ghisalberti & Nepf, 2002; Nezu & Sanjou, 2008). Momentum exchange between the two layers within and above the top of vegetation is controlled by the flow properties and to a large extent by vegetation form and stiffness (e.g. Wilson *et al.*, 2003).

6.4.3 Monami: A special case of turbulence

A characteristic and challenging phenomenon related to the interaction of flows with flexible vegetation is the wavy motion of the plants called *Honami* (Ikeda *et al.*, 2001) or *Monami* (Ghisalberti & Nepf, 2002). Ikeda & Kanazawa (1996) and Ghisalberti & Nepf (2002) showed that *monami* are due to the development of Kelvin–Helmholtz instability at the inflection point of the vertical profile of the mean velocity located at the tip of the vegetation. This instability leads to the generation of large, coherent vortices within the mixing layer, and the downstream progression of these vortices causes strong, periodic oscillations in the stream-wise velocity. As a result these vortices cause the progressive, coherent waving of the vegetation (*monami*). This has been confirmed by several studies (Ikeda & Kanazawa, 1996; Ikeda *et al.*, 2001; Ciraolo *et al.*, 2006; Yagci *et al.*, 2010) and is more than just a passive reflection of the flow structure. Ghisalberti & Nepf (2002) showed that when *monami* are present, the turbulent vertical transport of momentum is enhanced, with turbulent stresses penetrating an additional 30% of the plant height into the canopy.

6.5 THE IMPACTS OF PLANTS ON WAVES

Under the orbital motion of waves, plants move back and forth and therefore do not allow the onset of a skimming layer over a closed canopy as observed under unidirectional flow. Moreover, wave-induced motion varies on a much shorter time scale than unidirectional or tidal flow, which gives velocity profiles less time to develop (Denny, 1988).

6.5.1 Impacts on the velocity profile under waves

Profiles of wave energy spectra, analogous to velocity profiles under unidirectional flow, have been measured for crown-building aquatic species (e.g. *Amphibolis antarctica, Laminaria hyperborea*). In such species, where the majority of the biomass is located in the upper half of the canopy, a damping effect occurs within the canopy and maximum kinetic energy is just above the canopy (Verduin & Backhaus, 2000; Løvås & Tørum, 2001). However, in species where the biomass is evenly distributed throughout the canopy a steady decrease of orbital velocity with depth has been measured (Koch & Gust, 1999), suggesting an influence of plant morphology on vertical shear stress distribution (James & Barko, 2000).

Möller *et al.* (2011) claimed that stem areal density and height of vegetation affect wave transformation differently. While the number of stems per unit area controls the friction experienced by the horizontal flow components of the wave motion, the height of vegetation affects the distance of the boundary layer above the plants and thus the vertical component. Furthermore, and possibly more importantly, the physical

properties of the vegetation, such as its rigidity or flexibility, may also play a significant role in affecting wave transformations (see Section 6.3, this chapter).

6.5.2   Impacts on total wave energy

The effect of vegetation on wave motion is generally described by a reduction in
total wave energy which has been observed for salt marsh assemblages (Möller et al.,
1999; Bouma et al., 2010), kelp (Mork, 1996a) and sea grass (Fonseca & Cahalan,
1992). While previous studies agree that salt marshes are very effective in dissipating
wave energy (Wayne, 1976; Knutson et al., 1982; Möller et al., 1999; Cooper, 2005;
Bouma et al., 2010), observations are more variable for sea grass and kelp. Fonseca
and Cahalan (1992) found a reduction in wave energy density of ~40% per metre
of sea grass meadow in four different species and a reduction of significant wave
height of 50–85% was observed for the kelp *Laminaria hyperborea* (Mork, 1996a;
Dubi & Tørum, 1997). On the other hand, studies on the kelp *Macrocystis pyrifera*
did not find significant wave dissipation (Elwany et al., 1995). While these conflict-
ing results may be due to differences in the hydrodynamic conditions between studies, they show that it is still very difficult to generalise the effects of vegetation on
waves.

Similar to unidirectional flows, wave attenuation depends on the plant characteristics (see Section 6.3, this chapter), as well as on wave parameters (Méndez & Losada,
2004). Koch et al. (2009) suggest that the relationship between vegetation parameters
and wave parameters is non-linear, while Newell & Koch (2004) add that it may be
unsteady. Aquatic vegetation is exposed to wave spectra which are a combination of
a wide range of wave conditions including ocean swell, wind-generated waves and
boat wakes (Knutson et al., 1982; Koch, 2002; Ciavola, 2005; Paul & Amos, 2011).
In order to predict dissipation of spectral wave energy, it is important to understand
how vegetation interacts with waves of different frequencies. This is of particular
interest for flexible vegetation that may respond differently to different frequencies
(Augustin et al., 2009; Bradley & Houser, 2009; Manca, 2010). Bradley & Houser
(2009) suggest that if the natural frequency of, for example, sea grass is matched by
that of waves, then those waves will not be attenuated by the vegetation. Such observations indicate that the common simplification of vegetation elements in modelling
studies as rigid cylinders of varying diameter, spacing and orientation within a flow
field may be inadequate.

6.6   THE IMPACTS OF PLANTS ON THE SEDIMENTARY
ENVIRONMENT

The impact of vegetation on hydrodynamics has obvious corollaries for sediment
dynamics and erosion/deposition processes, as has been shown for submerged and
emergent aquatic macrophytes (Schulz et al., 2003; Cotton et al., 2006; Gurnell et al.,
2006; Asaeda et al., 2010; Pollen-Bankhead et al., 2011), kelp (Løvås & Tørum,
2001), reed (Jordanova & James, 2003; Türker et al., 2006; Rominger et al., 2010),
salt marsh assemblages (French et al., 2000; Callaghan et al., 2010) and pioneer
riparian tree species (Tooth & Nanson, 1999, 2000; Bertoldi et al., 2009; Corenblit

et al., 2009; Gurnell *et al.*, 2012). Vegetation exerts a number of controls on the geomorphological processes affecting the sedimentary environment. The manifestation of these controls are a series of hydraulic, hydrologic and mechanical effects, some of which have positive and some of which have negative impacts but all of which must be considered in experimental design, particularly if using live surrogates (see Chapter 3, this book).

The roots of plants are anchored in the substrate to support the above-ground parts of vegetation, thereby creating a reinforced matrix in which stress is transferred from the sediment to the roots, increasing the overall strength of the matrix (Greenway, 1987). The strength of a rooted substrate is, therefore, a combination of sediment and root strength combined with the strength of the bonds between sediment and roots (Waldron, 1977; Waldron & Dakessian, 1981; Ennos, 1990). Greenway (1987) notes that the magnitude of root reinforcement is a function of a number of factors:

- Root density
- Root tensile strength
- Root tensile modulus
- Root length/diameter ratio
- Sediment/root bond strength
- Alignment – angularity/straightness of the roots
- Orientation of the roots relative to the direction of principal strains

The extent to which root reinforcement is important for the stability of sediment in general and river banks in particular is also related to the sediment character, the size and dynamics of the river and the developmental stage and type of the vegetation (Baptist *et al.*, 2004; Hicks *et al.*, 2008). The ratio of bank height to rooting depth is particularly important in determining which types of vegetation will reinforce a given bank. These relationships are valid for dykes and banks developed to protect coastal areas.

Plant roots can influence the hydraulic scour of surficial material. In recent years, a significant amount of work has been undertaken to study the hydraulic effects of vegetation on geomorphic processes. Several laboratory flume and field studies have examined the effects of plant roots on upland concentrated flows (Mamo & Bubenzer, 2001a, 2001b; Gyssels & Poesen, 2003; Gyssels *et al.*, 2005; Zhou & Shangguan, 2005; De Baets *et al.*, 2006, 2007), and have shown an exponential decline in rill erodibility and soil detachment rates with increasing root-length densities and root biomass. Flume studies have shown that root architecture can play an important role in reducing soil erosion, with fine-rooted grasses being particularly effective at preventing soil detachment (De Baets *et al.*, 2006). The results of these studies are also relevant for streambanks lining incised channels or salt marshes and mangroves lining the littoral zone, with exposed root zones near to the water surface, because generally it is the root zone rather than the plant canopy that interacts with any flowing water. Exposed roots interact with flow like other vegetation parts (i.e. stems and leaves). The result of these interactions with transported sediment is the trapping and separation of fine sediment (Lowrance *et al.*, 1988; French *et al.*, 1995; Tsujimoto, 1999; Voulgaris & Meyers, 2004).

6.6.1   Sedimentation around and within plant stands

Movement of suspended matter is driven by the following processes: 1) downward settling due to gravity; 2) turbulent diffusion; and 3) advection by currents (usually parallel to the bed, but also perpendicular to the bed if a component of the velocity acts in that direction). Shi *et al.* (1995, 1996) and Braskerud (2001) proposed that turbulent wakes shed from stems may promote inter-particle collisions, accelerate flocculation and thus enhance settling of cohesive matter. Modified floc characteristics were found within the near-bed section affected by *Spartina anglica* (Graham & Manning, 2007) and where secondary maxima in both flow and shear stress occur (Leonard & Luther, 1995; Nepf & Vivoni, 2000). Turbulent diffusion is directed from regions of higher Suspended Sediment Concentration (SSC), usually found near the bed, towards regions of lower SSC, usually found higher in the water column, and its strength is proportional to the local suspended sediment concentration gradient and the turbulence intensity (Neumeier, 2007 and references therein). Turbulent diffusion and advection may therefore partially compensate for downward settling; within a dense canopy, downward particle motion is accelerated because of reduced turbulence, enhancing deposition of sediment. Low turbulence levels also favour trapping of sediment particles on leaves which, however, has been shown to account for a very small proportion (2–5%) of total deposition (French *et al.*, 1995). These findings agree with those of Ward *et al.* (1984) who found that sediment re-suspension by waves was suppressed and deposition slightly enhanced in vegetated regions in a shallow estuarine embayment.

Sharpe & James (2006) found that emergent vegetation promotes deposition of sediment from suspension by reducing flow velocity and vertical sediment diffusivity. In their experiments, stems reduced tranverse diffusivity in the shear zone along the interface between vegetated and non-vegetated areas and general vertical sediment diffusivity by approximately an order of magnitude. The extent of deposits within the stand of vegetation increased with flow depth and stem areal density and decreased with sediment grain size. Once deposited, the roots of vegetation stabilise deposits and accelerate their lateral and vertical growth (Baptist *et al.*, 2003; Crosato, 2008). French *et al.* (1995) observed that the rate of deposition decreased with distance from the edge of the vegetation and/or the transport conduit. Recent studies have shown that this decrease may be linked to enhanced turbulent kinetic energy between interacting stem wakes, precluding particle settling within the canopy and resulting in lower sedimentation compared to bare mudflats (Widdows *et al.*, 2008b) or sections at the edge of stands (Pratolongo *et al.*, 2010). Deposition and particle trapping has been studied extensively in sea grass fields (Ward *et al.*, 1984; Fonseca & Fisher, 1986; Gacia *et al.*, 1999; Koch, 1999; Terrados *et al.*, 2000; Vermaat *et al.*, 2000; Gacia & Duarte, 2001; Bos *et al.*, 2007; Hendriks *et al.*, 2008), salt marshes (Christiansen *et al.*, 2000; Temmerman *et al.*, 2005; Van Proosdij, 2006), mangrove forests (Mazda *et al.*, 1995; Wolanski, 1995; Furukawa *et al.*, 1997), and river channels (Pollen-Bankhead *et al.*, 2011). Conversely, Neumeier (2007) observed lower deposition rates near the edge of vegetation and attributed this to two processes: 1) the development of skimming flow, producing an oblique-upward current in the upper half of the canopy that advects suspended matter upward; 2) the development of a highly turbulent shear layer penetrating up to 0.5 m into the interface between vegetated and non-vegetated areas, mixing the water column and suspended sediment.

Results from studies of sedimentation within vegetated patches have therefore generally been contradictory in terms of the sediment retention potential of plants. This is a consequence of the wide range of vegetation characteristics (plant morphotype, stem areal density, total planform area of the stand), the nature of the hydrodynamic forcing (waves/currents, angle of attack, distance of the stand from channels or the littoral zone) and the underlying substrate characteristics, making it either difficult or impossible to isolate individual effects. These different results underline the need for careful experimental design in any physical ecohydraulic experiments incorporating live plants.

6.6.2   Scour and erosion

The effects of isolated individual plants and of densely-packed stands on localised scour and erosion are often disparate. Isolated or sparsely distributed plants locally reduce velocity but enhance turbulence (Fonseca & Koehl, 2006). As flowing water approaches an individual plant stem, water piles up on the upstream side of the stem and flow is accelerated around it, causing the formation of horseshoe vortices near the bed (*cf*. Richardson & Davis, 2001). As bed material is progressively scoured, the strength of horseshoe vortices reduce and eventually equilibrium is reached and scouring ceases. In addition vertical vortices, called wake vortices, form downstream of the stem (Figure 6.6). Both horseshoe and wake vortices accelerate the removal of material from around the base of the stem. As a result, sediment movement can be observed at velocities below the general threshold of motion, locally producing coarser sediment surfaces within sparse stands of vegetation (Fonseca & Koehl, 2006; Lefebvre *et al.*, 2010).

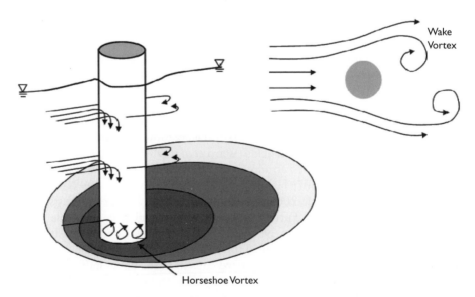

Figure 6.6 Schematic representation of scour around a plant stem (modified from a similar diagram for bridge piers by Richardson and Davis, 2001).

Erosion can be reduced by more complex, widespread and mixed plant stands (Brown, 1998; Neumeier & Ciavola, 2004). Salt marshes, for instance, are increasingly being advocated as viable mechanisms for "natural" coastal defence (Pethick & Burd, 1993), as long as they have the ability to maintain relative elevation within the tidal frame. This relies, in part, upon the accumulation of suspended particulate matter at a rate sufficient to keep pace with relative sea level rise (Graham & Manning, 2007). Thompson et al. (2004) concluded that the presence of sea grasses in the Venice Lagoon decreases bed erosion due to i) boundary shear stress reduction by ~20% and ii) stabilisation of the bed. These findings have been confirmed by other studies (French et al., 2000; Neumeier & Amos, 2006). Reduction of coastal erosion in the presence of *Posidonia oceanica* was also observed in the NW Mediterranean (Gacia, 2001). Similarly, coastal sub-littoral vegetation has been investigated with respect to its ability to stabilise beach profiles. The approach of using sea grass as soft beach erosion protection has been considered for decades. For example, Price et al. (1968) investigated the effect of artificial sea grass on beach erosion and concluded that vegetation can build up beaches by promoting onshore transport of material. Coops et al. (1996) found that the reed *Phragmites australis* had a greater impact on reducing wave erosion than *Scirpus* marsh. They reported strong wave attenuation through the stand, as well as significant reinforcement of the substrate by the network of rhizomes and roots (Knutson, 1988; Mallik & Rasid, 1993). Türker et al. (2006) found that i) a transition between dynamic and static stability can be established for beach profiles under the protection of emergent vegetation; ii) both the area of beach experiencing erosion and the area of an offshore bar decreased as the vegetated area increased; iii) damage to the beach was controlled by the dissipation of wave energy within the vegetation, and iv) in agreement with Möller (2006), flow depth was the most important factor controlling erosion. On the other hand, Elwany & Flick (1996) investigated the potential of offshore giant kelp beds (*Macrocystis pyrifera*) to influence the width of adjacent beaches on the Californian coast and found no clear correlation or consistent pattern. Similarly, Løvås & Tørum (2001) found that *Laminaria hyperborea* stands affected wave propagation during laboratory experiments, but had only weak control on beach or dune erosion.

6.7   NUMERICAL MODELLING OF THE INTERACTION BETWEEN VEGETATION, MEAN FLOW AND TURBULENCE

There is great strength in an approach to ecohydraulic research that combines numerical with physical modelling. As a result a brief discussion of developments in numerical modelling is included here.

Numerical models to predict the flow through vegetation with rigid and/or flexible stems have existed for several decades. Initial attempts to model the interaction of vegetation with water flows simulated the effects of submerged seaweed as a highly viscous layer (Price et al., 1968). Mork (1996b) extended the idea of the highly viscous layer and developed a theory for kelp that took into account not only viscous drag, but form drag for the canopy layer and the lower vegetative area. Vegetation has also been modelled as a high friction area by Camfield (1977) who studied wind-wave growth over shallow flooded regions. Asano *et al.* (1992) extended the conservation of momentum approach of Kobayashi *et al.* (1993) to include the movement of vegetation by coupling the flow field and swaying plant motion. Mendez *et al.* (1999) extended the previous momentum-based wave damping approach to include random waves. Lowe *et al.* (2005) developed a theoretical model that estimates the flow inside a submerged canopy under waves and predicted that in-canopy flow will be higher under oscillatory flow than under a unidirectional flow of the same magnitude. This model was subsequently extended to predict wave energy attenuation and its dependence on wave frequency (Lowe *et al.*, 2007). Both versions of the model have been validated using an idealised system of rigid vertical cylinders. Originally, the cylinder dimensions were chosen to represent coral canopies, but a recent study with artificial sea grass surrogates suggests that it may also be at least partly applicable to flexible vegetation (Manca, 2010). All these models linearise the drag force acting on the plant surrogate; they have been developed for horizontal substrates without incorporating breaking waves and do not seek to explicitly simulate the motion of plants. Løvås (2000) extended the Larson (1995) model to take into account the effect of kelp, including variable depth and wave breaking.

To simulate the feedback between turbulent flows and the flexure of vegetation, it is critical that turbulent fluctuations are adequately simulated. Turbulence models range in complexity from simplified velocity profile approximations, through Reynolds- and Double-Averaged Navier-Stokes (RANS and DANS, respectively) equation solvers, to Large Eddy Simulation (LES) models and most recently Direct Numerical Simulation (DNS). Potentially also of interest are more recent efforts to predict open-channel flow using data-driven approaches such as artificial neural networks (Abdeen, 2008; Wu *et al.*, 2009).

The classical approach to close the Navier-Stokes (momentum) equations is by Reynolds decomposition of the velocity components into time-averaged and fluctuating components, and then to link the resulting Reynolds stresses to properties of the time-averaged flow. This is known as the Reynolds-averaged Navier-Stokes (RANS) approach. RANS models are computationally efficient since turbulence fluctuations are modelled by empirical equations and a coarse grid, as well as large time steps, can be used. However, the RANS approach focuses upon accurately representing the mean flow field to the detriment of velocity fluctuations and hence coherent flow structures (Keylock *et al.*, 2005). In principle, it is possible to solve the Navier-Stokes equations using DNS (Moin & Mahesh, 1998), but current computing technology limits the application of DNS to only marginally turbulent flows. LES has been developed extensively in the last 15 years in the turbulence and fluid engineering communities and can be viewed as an intermediate case between DNS and RANS approaches (Keylock *et al.*, 2005). While DNS deals with all eddies larger than the smallest (dissipation) scale and RANS methods deal with the mean flow characteristics, LES calculates the properties of all eddies larger than a filter size and models those smaller than this scale by a subgrid-scale turbulence transport model (Keylock *et al.*, 2005). Vegetation-generated turbulence is commonly either included as a source term in the sub-grid scale model (Su & Li, 2002), or by the immersed boundary method (Stoesser *et al.*, 2009). In open channel flow with submerged vegetation, turbulence is predominately generated by the shear discontinuity at the interface of the vegetated and non-vegetated region while the stem-generated turbulence is comparatively small and may be neglected (Cui & Neary, 2008). Patton *et al.* (1998) used the one-equation k-l sub-grid turbulence closure, while other researchers used the Smargorinsky sub-grid scale turbulence closure with dynamic adjustment of the closure coefficient (Cui & Neary, 2008; Li & Yu, 2010). Recently, hybrid LES/RANS models have been developed (e.g. Li & Yu, 2010) that simulate simpler flows at the larger extent of the domain using a RANS module and use LES only for vegetated regions where complicated flows occur.

The deflection of flexible vegetation under loading has many similarities with the bending of an elastic beam. The bending and waving of plants then lends itself to analysis by Timoshenko's beam theory (Timoshenko, 1955). This theory is limited to small deflections of a cantilevered beam and assumes prismatic stems of relatively high stiffness. It has been applied in one dimension (1D, Kutija & Hong, 1996) and extended to a quasi-three-dimensional (3D) method by Erduarn & Kutija (2003). Velasco *et al.* (2008) developed a 1D model computing vertical velocity and Reynolds stress profiles, which employed the classical elastic beam equation for the deflection of vegetation stems with moderate flexibility. Abdelrhman (2007) developed and tested a two-dimensional (2D) model, which modelled a blade as a series of elements, for coupling flow and very flexible eelgrass. In a similar approach, Dijkstra & Uittenbogaard (2010) incorporated plants consisting of a number of leaf segments in a 1DV *k-ε* turbulence model to model very flexible vegetation. The model is complemented by another study that focuses on the effect of unidirectional flow on vegetation posture and drag taking both plant stiffness and buoyancy into account (Luhar & Nepf, 2011). Ikeda *et al.* (2001) developed a sophisticated 2D LES model to simulate the wavy motion of flexible vegetation. Their model directly solved the equation of motion of each flexible stem and used a complex "plant grid" to track the movement of each stem. Li & Yan (2007) developed a fully 3D numerical model, which can also simulate wave-current-vegetation interaction phenomena. This model utilised the split-operator approach, in which the advection, diffusion, and pressure propagation are solved separately. The unsteady fluid force on vegetation was split into a time-dependent inertial component and a drag component, and vegetation was modelled as a sink of momentum. More recently and using a similar approach, Li & Zhang (2010) presented a 3D RANS model to investigate the hydrodynamics and mixing induced by random waves acting upon vegetation. The model used the Spalart–Allmaras model for turbulence closure. Li & Xie (2011) developed a 3D numerical model of the hydrodynamics of submerged flexible vegetation with or without foliage. Flexible vegetation was modelled using momentum sink terms, with the velocity-dependent stem height determined by a large deflection analysis using the Euler-Bernoulli Law for the bending of a slender beam. This approach is more accurate than the small deflection analysis of Timoshenko (1955).

6.8 CONCLUSIONS

Significant effort has been devoted to developing a better understanding, and improving prediction, of the interactions between vegetation and water flows. Plants can act as ecosystem engineers and hence generate suitable conditions for themselves as well as other plant and animal species (Bouma *et al.*, 2005). This mainly results from the reduction of flow velocity and turbulence that results in sediment accretion or reduced erosion, but it can also be a function of the diversion flow, e.g. in braided channels. On the other hand, vegetation can increase flood hazards (i.e. on floodplains), if it blocks the passage of high discharge events. A larger volume of research exists for unidirectional flows compared with that for plant-wave interactions. This might be attributed to i) the maturity of hydraulic engineering as a discipline in comparison to coastal engineering, with more applications and a larger research community; ii) the important knowledge gaps still existing, leading scientists to focus on the more "fundamental" unidirectional flow case; iii) the fact that unidirectional flows allow more time averaging and thus require less intensive computations for numerical modelling. Relatively few studies have addressed the effect of vegetation on combined waves and currents. There have been two notable recent exceptions. Gaylord *et al.* (2003) investigated the effect of an alongshore current on wave dissipation by kelp and Paul *et al.* (2012) found that the wave attenuating capacity of sea grass was reduced in the presence of a current following waves. As a result, wave-current-plant interaction processes are largely unknown and constitute an important direction of future research and further scientific progress. For example, previous laboratory studies have discussed the effect of plant density and configuration, both relative to individual stems and to the flume lateral walls in unidirectional flows, while such knowledge is limited for waves. Furthermore, wave attenuation by plants, not only horizontally but also vertically resulting in bed shear stress reduction, still remains poorly understood compared with processes under unidirectional flows. Physical modelling in hydraulic laboratories has an important role to play in filling these gaps in knowledge. However this chapter has highlighted the complexity of plant-flow interactions and the need for care in the design and implementation of such experiments. Careful selection of the facility, plants, sediments, flow and both measurement methods and sampling frequency are central to successful outcomes and good data. It is evident that there is a great need for well-designed hydraulic laboratory experiments to explore more fully plant-flow interactions.

Chapter 7

Macrozoobenthos, hydraulics and sediment dynamics

M.F. Johnson & S.P. Rice

7.1 INTRODUCTION

This chapter reviews the impacts of mobile and sessile invertebrate macrozoobenthos on hydraulic and sedimentary processes in aquatic environments, emphasising the factors that should be considered when carrying out hydraulic experiments. Here, macrozoobenthos includes animals that greater than 1 mm in length and, therefore, does not include micro- or meio-fauna and the important stabilising effects of their secretions (Wotton, 2011). It also does not include large communities of micro- and meio-fauna, such as corals and bryozoans. In addition, vertebrates including fish, are deliberately excluded from this review; relevant reviews of the interactions between fish, hydraulics and sediment dynamics can be found in Atkinson and Taylor (1991), Butler (1995), Hassan *et al.* (2008) and Rice *et al.* (2012), among others. Sessile animals are those that live attached to the solid-water interface whereas mobile animals move freely within or on sediments.

Animals modify the hydraulic environment as currents adjust to the presence of their bodies, increasing heterogeneity by creating areas of accelerating and decelerating flow and inducing drag and lift forces (section 7.2). In addition, many animals suspension-feed which can involve generating currents to transport suspended organic matter to feeding appendages and mixing the water column to prevent seston depletion near the bed (section 7.3). Animals also mix sediments when burrowing and feeding, altering the biochemistry, structure and topography of the substrate (section 7.4). This modification of the bed surface has implications for near-bed hydrodynamics (section 7.5) and, as a result animals may also have substantial impacts on sediment stability (section 7.6). Where animals occur in large aggregations, their environmental impacts can be substantial and persistent, creating biogenic habitats that support large and diverse assemblages of organisms which could not otherwise survive: that is, they can be effective ecosystem engineers.

7.2 DRAG ON AND AROUND ANIMALS

7.2.1 The significance of drag for animals

Most sessile animals suspension-feed and, consequently, are reliant upon water currents to deliver food and oxygen and to remove waste products and filtered water. Flowing water is also important for the dissemination of eggs, sperm and larvae. Consequently, sessile organisms need to balance the positive impacts of flow with the negative aspects of flow stress. This stress is manifest as "added mass", buoyant, lift and drag forces. Drag is the force predominantly associated with the pressure difference between the upstream and downstream sides of an object. Its impact on animals has received much attention in the literature. For instance, the drag on blackfly larvae in streams was measured as 2.7×10^{-4} N at a velocity of 0.1 m s^{-1} and 4.24×10^{-4} at a velocity of 0.9 m s^{-1} but doubled when the larvae extended labral fans to filter feed (Eymann, 1989). Consequently, drag is an important stressor of sessile animals and must be minimised or mitigated to prevent dislodgement.

7.2.2   Variations in the drag acting on sessile animals

Drag is related to the velocity of the fluid relative to that of the animal, as well as the fluid viscosity and density and hence the Reynolds number. In addition, the drag acting on an animal is a function of the geometric and material properties of the organism and its exposed surfaces, including the size, shape, surface area to volume ratio, roughness and elasticity. Consequently, it is partially determined by:

- Animal morphology
- The surface texture of the animal
- Animal behaviour
- The presence of organisms attached to the animal.

Experiments studying the interaction between animals and flowing water therefore need to carefully consider the selection of organisms in order to ensure replicability between experiments. In addition, the behaviour of animals during experiments will influence results and, therefore, it is important to provide experimental conditions that allow animals to behave in a manner similar to field equivalents. This is especially the case if experimental measurements are to be used to infer behaviour under field conditions. See Chapters 2 and 3 for fuller consideration of these issues.

7.2.2.1   Animal morphology

Body morphology frequently reflects a balance between minimising drag whilst maintaining feeding capability. For instance, many organisms (goose barnacles, sea anemones, black fly larvae, polychaete worms) have a streamlined base that adheres strongly to a surface with an arm or tentacles that extend into the overlying, faster flow layers (Vogel, 1994). However flow is not always the most important factor controlling morphology since there are many other selection pressures including predation, competition and calcium availability (Kemp & Bertness, 1984; Rundle et al., 2004).

7.2.2.2   Animal texture

Surface texture modifies the boundary layer around an animal. Some fluid dynamical studies have suggested that the rough shell texture of the limpet Scutellastra argenvillei creates a turbulent boundary layer, reducing its wake by preventing flow separation and thus reducing drag (e.g. Branch & Marsh, 1978). Consistent with this, Beaumont & Wei (1991) found sub-littoral limpets (Nacella polaris) had smoother shells than their littoral counterparts, presumably because drag was more important due to strong tidal currents. In addition, some qualitative studies have suggested that textural shell ornamentation in terms of ribs and postules can reduce drag (Stanley, 1981; Watters, 1994). While increased surface roughness may conceivably reduce form drag by increasing boundary turbulence and reducing streamwise pressure differences over the object, this is likely to be more than compensated for by the increase in skin friction. As a result, a single, straightforward, relationship between shell texture and total drag is unlikely to hold true for most animals.

7.2.2.3 Animal behaviour

Behavioural changes can minimise drag forces on animals. The sea anemone *Metridium senile* has a complex response to increasing flow rates and can reduce drag forces by reconfiguring its tentacles or retracting them completely (Figure 7.1) (Koehl, 1977). Branch & Marsh (1978) found that drag on front-facing limpets (*Patella* spp.) was lower than that associated with side- or back-facing orientations and Warburton (1976) found that limpets (*Patella pellucida*) were preferentially orientated parallel to the flow in currents faster than 0.5 m s^{-1}. Similarly, García-March *et al.* (2007) reported that mussels that were orientated laterally had drag forces 400% greater than those orientated dorso-ventrally. Mobile animals have also been found to reorient their bodies to minimise drag forces. For example, Maude & Williams (1983) found that eight species of crayfish postured in flowing water to streamline their body form.

7.2.2.4 Attached organisms (epibionts)

Sessile animals, such as mussels, are often the only hard substrate available to other organisms, particularly in marine environments, leading to the colonisation of sessile animals by epiphytes and epizoans (attached plants and animals, respectively). These organisms (epibionts) increase the area exposed to the flow, thereby increasing the drag forces operating on the colonised organism. The height of the largest fouling organism and the surface distribution has the dominant influence on drag (Schultz,

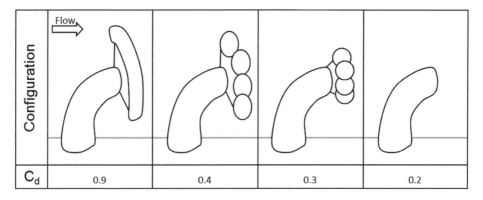

Figure 7.1 The changing morphology of the sea anemone Metridium senile with increasing flow velocity from left to right leading to a reduction in drag coefficient (C$_D$). Modified from Koehl (1977).

2004). For a constant flow rate and surface texture, drag increases with the logarithm of "roughness density" (RD), defined as the ratio of organism area to sampling area and expressed as a percentage, such that a RD of 5% causes 66% of the increase in drag that is caused by a RD of 75% (Kempf, 1937; Townsin, 2003; Schultz, 2004).

Witman & Suchanek (1984) used a force transducer to measure the effect of colonisation of *Mytilus edulis* by kelp (*Saccharina latissima*) and brown algae (*Alaria* spp.). Not only did kelp increase the drag force by 600%, but it also modified the orientation of mussels, elevating and twisting them and increasing the projected area of their shells by 278–297% (Witman & Suchanek, 1984). Periwinkles (*Littorina littorea*) colonised by various epibionts all suffered an increase in drag at current velocities of 80 mm s^{-1} (Wahl, 1996; Figure 7.2). The presence of some epibionts,

Host species	Epibiont species	% increase in volume	Impact	Reference
Periwinkle (*Littorina littorea*) 	 Barnacle (*Amphibalanus improvisus*)	44.9	160% increase in drag	Wahl (1996)
	 Filamentous brown alga (*Ectocarpus* sp)	24.8	220% increase in drag	Wahl (1996)
	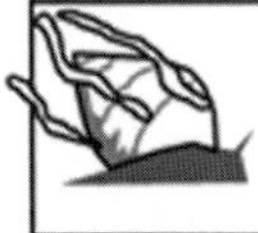 Green alga (*Ulva intestinalis*)	1.9 – 4.4	Negligible change in drag	Wahl (1996)
	 Oyster (*Crassostrea gigas*)		Large scale stranding	Eschweiler & Buschbaum (2011)
Scallop (*Chlamys hastata*) 	 Sponge		C_D decreased from 0.9 to 0.7	Donovan et al. (2003)
	 Barnacle (*Balanus* spp.)		C_D increased from 0.7 to 1.0	Donovan et al. (2003)
Mussel (*Mytilus edulis*) 	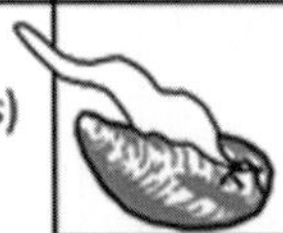 Kelp (*Saccharina latissima; Alaria marginata*)		200 – 600% increase in drag	Witman & Suchanek (1984)
	 Slipper limpet (*Crepidula fornicata*)		Increased attachment thread production	Thieltges & Buschbaum (2006)

Figure 7.2 The impact of epibionts on host molluscs, with an indication of the impact of colonisation.

such as barnacles and sponges, also increases the weight of host organisms and thus may counteract buoyancy and drag in the dislodgement of organisms, but will potentially inhibit other ecological processes such as locomotion. It is therefore essential to consider whether, based on the aims of the research, epibionts should be present on animals used in experiments.

7.2.3   Measuring drag on sessile animals

Drag forces can be estimated from measurements of the pressure difference over an animal (Denny, 2000), but it is more usual to measure it directly using force transducers (Chapter 4). The animal, such as a mollusc, is attached to a solid plate that is fixed to the transducer. Strain gauges in the transducer are deformed by the horizontal, streamwise, force (drag) acting on the object attached to the plate and the magnitude of deformation is directly related to the drag force. The transducers must be calibrated before use, usually by measuring known forces using a series of weights. Those used by Witman & Suchanek (1984) to measure drag on mussels had a standard error of 2.5%. Branch & Marsh (1978) found the error was greatest at low velocities or when using small shells, probably because the measured drag force was too small to register accurately. Most previous studies have been undertaken in flumes, but some drag measurements have been made in the field. O'Donnell (2008) quantified the wave dampening impact of bivalves by measuring the drag on a small sphere (9.5 mm diameter) that was placed in the middle of a mussel bed and connected to a force transducer with nylon thread.

The magnitude of the drag force is dependent on the flow conditions, particularly velocity. In order to compare between organisms, and under different hydrodynamic conditions, the dimensionless drag coefficient (C_D) is often used. However, the calculation of C_D is dependent on many biological and experimental factors that add variability to recorded measures. For instance, Cooper et al. (2007) highlight the significance of the ratio of object wetted perimeter to the flume cross section (blockage ratio) for drag measurements. They measured the drag over cylinders at the same flow velocity in two flumes, one 0.25 m wide and the other 0.75 m wide. The difference in blockage ratio between flumes resulted in discrepancies in the measured drag of 2.5 to 7.5% for rigid cylinders and up to 22% when using submerged macrophytes. Consequently, experiments focusing on the drag acting on organisms need to be carefully designed and quoted values of C_D should be assessed critically before making comparisons between studies (see Chapter 4 for further discussion).

7.3   FLOW GENERATION AND BIOIRRIGATION OF SEDIMENTS

7.3.1   The environmental significance of flows generated by animals

Suspension-feeding, sessile animals are not inanimate roughness elements. They draw in water through an inhalant siphon, pump it over their gills to filter out organic matter, and expel it through an exhalant siphon (Jørgensen et al., 1986; Wildish & Kristmanson, 1997). Clams and mussels direct exhalant jets vertically away from themselves whereas oysters direct exhalant siphonal flow horizontally (Troost *et al.*, 2009). Jetting of exhalant water prevents filtered water from being re-filtered by the animal, especially when current velocity is low relative to siphonal jet velocity (Fréchette *et al.*, 1989; Monismith *et al.*, 1990; Butman *et al.*, 1994; Jones *et al.*, 2011). Thus, these jets can mix near-bottom water with the overlying flow (Ertman & Jumars, 1988; Larsen & Riisgård, 1997). A compilation of siphonal velocities generated by bivalves and ascidians is given in Table 7.1. Monismith *et al.* (1990) found that the disturbance of the boundary layer induced by exhalant siphons extended 0.1 m (30 diameters) downstream of surrogate *Venerupis philippinarum*. André *et al.* (1993) also detected the excurrent flow of *Cerastoderma edule* 0.1 m downstream of siphons but the vertical extent was only 0.01 m above the bed at ambient velocities of 0.15 m s^{-1}. Due to the relatively strong velocity of exhalent jets, oncoming flow slows and is forced around the exhalent jet, modifying both mean flow and turbulence properties and acting as an additional source of roughness (Figure 7.3) (Eckman & Nowell, 1984; Monismith *et al.*, 1990; O'Riordan *et al.*, 1993; Lu *et al.*, 2000; van Duren *et al.*, 2006). For instance, *C. edule* was estimated to increase z_0 by 286% and 4681% at 0.05 m s^{-1} and 0.4 m s^{-1}, respectively, using the law of the wall (Fernandes *et al.*, 2007). This was attributed to siphonal currents because the increase in z_0 did not coincide with a conspicuous visual increase of physical bed roughness (Fernandes *et al.*, 2007). However, other studies have found siphons to be of limited significance relative to the roughness generated by the body of animals (Butman *et al.*, 1994; Plew *et al.*, 2009). It should also be noted that in natural populations of animals, siphoning occurs only some of the time, related to tidal cycles and food availability (Saurel *et al.*, 2007).

Many animals also actively pump water from a burrow in order to filter organic matter, eject sediment when burrowing, and remove waste products. This process is termed bioirrigation. In addition to active filtering, the mound and funnel/pit topography of many burrow structures (see section 7.4) creates pressure differences as the flow adjusts to the altered topography, driving pore-water exchange and passive irrigation of burrows (Allanson *et al.*, 1992; Huettel & Gust, 1992; Yager *et al.*, 1993; Ziebis *et al.*, 1996). For instance, lugworms (*Arenicola* spp.) build a burrow with a shallow pit at one opening and a conical faecal mound at the other. The anterior end of the worm faces the pit and, thus, whenever water is moving, there is passive flow from the anterior to the posterior end of the worm (Vogel & Bretz, 1971). When pumping water from burrows, animals lower the pore-water pressure, causing the movement of overlying water into the sediment, sometimes referred to as bioadvection (Woodin *et al.*, 2010). Bioadvection is similar to the purely physical exchange between pore-water and oxygenated surface waters, but the latter is usually confined to the upper 50 mm of sediments, whereas animals can mix water hundreds of millimetres deep (Meysman *et al.*, 2007; Volkenborn *et al.*, 2010; Woodin *et al.*, 2010). Exchanges between pore-water, burrow-water and the overlying water column have implications for biogeochemical processes including the oxygenation of sediments and the transformation, exchange and flux of nutrients and trace elements (Matisoff *et al.*, 1985; Nates & Felder, 1998; Kristensen & Hansen, 1999).

Aggregations of animals, such as sponges, bivalves, polychaete worms and ascidians, generate population filtration rates typically in the range of 1 to 10 m^3 water m^{-2} day^{-1} (Riisgård & Larsen, 2000). Consequently, communities of animals

Table 7.1 Siphonal jet velocities in bivalve and ascidian species.

Species	Taxa	Size (mm)	Exhalant (mm s⁻¹)	Ref.
Chironomus plumosus	Midge larvae (insect)	20	15	Roskosch et al. (2010)
Nereididae spp.	Worm (Polychaete)		10	Larsen & Riisgård (1997)
Clinocardium nuttallii	Cockle (bivalve)		90–110	Ertman & Jumars (1988)
Cerastoderma edule	Cockle (bivalve)		208	Troost et al. (2009)
Argopecten irradians	Scallop (bivalve)		107 ± 22	Frank et al. (2008)
Mercenaria mercenaria	Clam (bivalve)		240 ± 14	
Anodonta spp	Mussel (bivalve)	15	106–143	Price & Schiebe (1978)
Mytilus spp.	Mussel (bivalve)	30–40	60	LaBarbera (1981)
Mytilus edulis	Mussel (bivalve)	23–24		Green et al. (2003)
Mytilus edulis	Mussel (bivalve)		432 ± 039	Frank et al. (2008)
Mytilus edulis	Mussel (bivalve)	68	185	Troost et al. (2009)
Crassostrea gigas	Oyster (bivalve)		243–486	
Crassostrea virginica	Oyster (bivalve)		14 ± 2.2	Frank et al. (2008)
Styela clava	Sea squirt (ascidian)		2.7 ± 1.0	Frank et al. (2008)
Styela plicata	Sea squirt (ascidian)	70	530–107	Fiala-Médiona (1978a;
Ascidia mentula	Sea squirt (ascidian)	80–160	34	1978b)
Ciona intestinalis	Sea squirt (ascidian)	65–75	57–95	
Microcosmus sabatieri	Sea squirt (ascidian)	90–120	52–194	
Phallusia mammillata	Sea squirt (ascidian)	120–150	39	

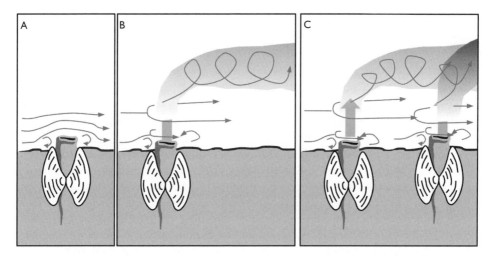

Figure 7.3 Siphonal feeding currents of clams. A: flow acceleration over siphons. B: flow with active siphoning and C: complex flow with multiple siphons (after O'Riordan et al., 1995).

can filter large volumes of water, often measured to be several times the volume of the overlying water column, in a day (Jørgensen, 1990; Petersen & Riisgård, 1992; Roditi et al., 1996; Riisgård, 1998; Dolmer, 2000b; Forster & Zettler, 2004). In the River Spree, Germany, communities of Unionid mussels (*Anodonta anatina*, *Unio crassus*, and *U. pictorum*) and zebra mussels (*Dreissena polymorpha*) are estimated to filter a volume of water equal to the mean daily stream discharge (320,000 m³ day⁻¹) (Welker & Walz, 1998). Similarly, the shrimp *Upogebia pusilla* is estimated to pump the same volume of water through their burrows in 24 hours as the neap tide in the lagoon of Grabo on the Adriatic coast (Dworschak, 1981). While this has obvious ecological consequences, the physical consequences for boundary layer hydraulics have yet to be examined. However, it seems likely that the periodic jetting of water through narrow burrows will increase near-bed turbulence, so temporal and spatial patterns of siphoning should be considered when designing laboratory experiments.

7.3.2 Factors affecting animal-generated flows

Siphonal flows can be highly variable between species, within species and for an individual over time. This variability is related to:

- Siphon geometry (size and orientation)
- Ambient flow velocity
- Seston concentration and quality.

These factors are considered below and have important implications for experiments where biogenic flows are present. Studies of biogenic roughness should also consider the potential importance of siphonal flows and burrow irrigation as additional roughness sources that will not be generated in adverse environmental conditions or mimicked by surrogate animals (Chapter 3, this book).

7.3.2.1 Siphonal geometry

The velocity of siphonal flows is related to the cross-sectional area of the siphon opening, which is variable between individuals and can alter during or between observations. For example, Fiala-Médioni (1978a) observed that the pumping rate of ascidians was fairly constant during individual flume experiments but was variable between experiments due to alterations in siphon diameter. Troost *et al.* (2009) also found that variations in siphonal flows between and within species of shellfish were related to variability in siphon diameter. The orientation of siphons relative to the flow direction also impacts upon the extent of hydraulic disturbance. Orientating siphons into the flow is hypothesised to reduce pumping costs (Englund & Heino, 1996) whereas orientating perpendicular to the flow reduces interactions between siphons, removing the threat of refiltration of exhalant water (Vincent *et al.*, 1988). Infaunal bivalves completely or partially bury themselves with their siphons flush with, or slightly above, the sediment surface. The height of the siphon above the bed increases surface roughness and turbulent mixing, reducing refiltration of water (O'Riordan *et al.*, 1995). Siphons have been observed extending well above the surface in *Venerupis* sp. (15 mm extension) and *Potamocorbula* sp. (5 mm extension) in a laboratory flume (Monismith *et al.*, 1990). Because of the influence of siphon position and orientation on near-bed hydrodynamics, animals introduced to laboratory flumes should be permitted to position themselves under the flow conditions that will be used in experiments (Chapter 2, this book). Failure to do so may result in data that are not analogous to nature.

7.3.2.2 Ambient flow conditions

Siphoning behaviour is dictated, in part, by flow velocity. Filtration rates of siphoning clams increased 300% over flows ranging from 0.15 to 0.25 m s⁻¹ (Cole *et al.*, 1992). Conversely, Wildish & Miyares (1990) found that filtration by mussels (*Mytilus edulis*) was inhibited as flow increased from 0.06 to 0.23 m s⁻¹. Sobral & Widdows (2000) also found that filtering by infaunal clams (*Venerupis decussata*) increased with water velocity up to 0.08 m s⁻¹ but then decreased, particularly above 0.17 m s⁻¹. Therefore, filtration by bivalves may increase over a range of flows to an upper limit, after which it is physiologically or hydrodynamically not feasible (Wildish *et al.*, 1987; Cole *et al.*, 1992; Grizzle *et al.*, 1992; Butman *et al.*, 1994). Similarly, it is known that the beating of fans by barnacles is related to flow velocity, with passive feeding at flows >20 mm s⁻¹ and active fan beating to draw in water to feeding appendages at velocities <20 mm s⁻¹ (Trager *et al.*, 1990; Hunt & Alexander, 1991). Consequently, consideration of how flow velocity will affect the activity of animals is essential when planning experiments on biogenic flows.

The significance of biogenically-generated flows is also relative to the ambient flow velocity. At velocities of 0.3 m s⁻¹, Widdows *et al.* (2009) found that mussels increased Turbulent Kinetic Energy (TKE) and bed shear stress by an order of magnitude in comparison to bare sediment and they suggested that 40% of this increase was due to active siphoning. Exhalant jets have also been found to increase TKE in flows of 40 and 80 mm s⁻¹ over *Mytilus edulis* beds (Lassen *et al.*, 2006). At velocities of 0.055 m s⁻¹, van Duren *et al.* (2006) found that TKE production due to mussel filtration was greater than that generated by bottom shear but was negligible at 0.35 m s⁻¹. The impact of siphoning activity is also limited to only a few centimetres above the bed (van Duren *et al.*, 2006; Crimaldi *et al.*, 2007). Similarly, Thomason *et al.* (1998), using flow visualisation in a 5 m long flume, found active feeding by barnacles increased the thickness of the boundary layer by generation of turbulence. However, this effect was flow dependent and was negligible at flow velocities in excess of 0.14 m s⁻¹. Therefore, the hydrodynamic impacts of biogenic flows are of biological significance at small scales and low velocities, but at larger scales and faster flows it is likely to be negligible.

7.3.2.3 Seston concentration and quality

Another major consideration for all studies of feeding currents should be the quantity and quality of seston. Bivalves have been found to siphon only when seston is present in the flow and, consequently, food should be provided for all filter feeders. Bivalves also adjust their filtration rates in response to fluctuations in seston concentration (Dolmer, 2000b; Riisgård *et al.*, 2003). For instance, Dolmer (2000a) observed that 44–69% of mussels remained closed (not siphoning) when phytoplankton was depleted near the bed whereas only 17% remained closed when phytoplankton was present at the bed.

7.3.3   Quantifying flows generated by animals

In order to identify and measure the velocity of feeding currents and their effects on the surrounding hydrodynamic environment, sensitive flow measurement devices are required. Particle Image Velocimetry (PIV) allows multi-point flow measurements with millimetre spacing over small areas, which would be ideal for this type of study. However, PIV requires seeding material to be tracked by cameras and actively filtering sessile animals can complicate this procedure because their feeding apparatus intercepts seeding material (Frank *et al.*, 2008; Troost *et al.*, 2009). The rate of filtration by bivalves is affected by the suspended load which is effectively increased by adding seeding material to the flow and thus may affect the filtering rate. In addition, sessile animals will actively remove seeding from the flow when filter-feeding so the material may need to be replenished during experiments. Issues also exist because excurrent jets will be depleted of seeding material due to filtration, reducing the amount available for flow tracking. Therefore, the use of flow measuring devices that require seeding material needs to be fully evaluated before they are used to study siphonal animals. In addition, seeding material can have detrimental impacts on animal health and behaviour when ingested or when present in quantities sufficient to smother organisms (see Chapter 2 and 4 for further details).

The irrigation of sediments by benthic animals has largely been quantified by identifying oxic halos around burrows in otherwise anoxic sediments, indicating that animals have oxygenated surrounding sediments by exchanging water. Direct measurements of the pumping rates of animals in burrows have been achieved using a number of methods. Riisgård (1991) placed polychaete worms (*Hediste diversicolor*) in small glass tubes sitting horizontally between two chambers and monitored the water level between chambers while worms pumped water between them. Similarly, Dworschak (1981) measured the water overflow from a basin containing a Thalassinidean shrimp in an artificial burrow. The results from these studies rely on the assumption that animals behave the same in artificial burrows as they would in natural ones and, given temporal variability in animal behaviour (seasonally or over much shorter periods), it is presently unknown whether short-term pumping observations provide reliable estimates of long-term rates. The volume of water pumped by Thalassinidean shrimp has been quantified *in situ* by Colin *et al.* (1986) by pushing funnels several centimetres into the sediment over active shrimp mounds such that pumping of water from the mound produced an equal displacement of water at the funnel opening which was measured.

7.4   SEDIMENT MIXING BY MOBILE ANIMALS (BIOTURBATION)

7.4.1   The significance of bioturbation for aquatic sediments

Many mobile animals burrow into substrates, vertically mixing layers of sediment. This process is termed bioturbation and has been studied for over 100 years in terrestrial environments (Darwin, 1881; Feller *et al.*, 2003). In fine-grained marine sediments that constitute 70% of the Earth's surface, bioturbation is ubiquitous. Bioturbation is thought to have been fundamental to the evolution of the marine biome (Thayer, 1979; Canfield & Farquhar, 2009). Fossil evidence suggests that increased burrowing by organisms at the Precambrian-Cambrian transition, approximately 540 million years ago, led to the reworking and oxygenation of the ocean floor (Thayer, 1979; Crimes & Drosser, 1992; Seilacher et al., 2005), enabling the evolution of many of the major groups of marine animals which today constitute the largest biomass of any environment on Earth (Bottjer et al., 2000).

Bioturbation is of biological significance because it creates a habitat that can be utilised by a greater diversity of organisms than sediments that are not mixed. The presence of burrows increases the surface area of the water-sediment interface, with implications for the efficiency of exchange between interstitial and overlying water. Fiddler crabs of the genus Uca have been recorded to increase this area by 59% (Katz, 1980) whereas King ragworms (Alitta virens) can cause an increase of 325% (Gerino, 1990). Burrow shape and wall architecture also control diffusion (Aller, 1988; Zorn et al., 2006). This has important implications for biogeochemistry by increasing oxygen penetration, stimulating microbial metabolism and cycling nutrients and solutes at the water-sediment interface (McCall et al., 1979; Matisoff et al., 1985; Levinton, 1995; Vaughn & Hakenkamp, 2001; Lohrer et al., 2004). Consequently, bioturbation is of crucial importance to benthic ecology (Biles et al., 2002). A meta-analysis by Ofalsson (2003) found that 86% of the 77 reviewed articles on macrofaunal bioturbation identified that the macrobenthos increased species diversity of meiofauna as a result of biogenic structures. Bioturbation by macrofauna also transports pollutants and can increase the degradation of organic pollutants. Banta & Andersen (2003) found that Hediste diversicolor actively metabolised organic pollutants, flushed dissolved contaminants from sediments and stimulated microbial degradation of contaminants. Bioturbation also influences the dispersal and degradation of hydrocarbons following oil contamination by altering oxic/anoxic boundaries in sediments, increasing microbial communities, and transporting sediments across these boundaries (Cuny et al., 2007).

Bioturbation can change sediment structure, roughness and stability. This is associated with local alterations in packing, porosity, compaction and rigidity in the vicinity of burrows (Aller & Yingst, 1980), which increase the spatial heterogeneity of sediment structures and can alter properties over large areas ($>10^1$ m^2). For instance, bioturbation by both oligocheate and polychaete worms decreases the compactness of sediment surfaces, increasing sediment porosity and water content (Fukuhara, 1987; Meadows & Tait, 1989) and decreasing sediment stability (section 7.5). If bioturbating organisms are removed from sediments using a biocide, these sediments become increasingly compacted and stable (Underwood & Paterson, 1993). The process is dependent on species-specific burrowing characteristics and, consequently, the type and magnitude of bioturbation is usually related to functional grouping (see Table 7.2; Griffis & Suchanek, 1991; Francois et al., 2002; Gerino et al., 2003).

Bioturbation by polychaete worms has been widely studied because they often dominate the macrofauna in temperate marine environments and have substantial mixing capabilities. For example, a 0.06–0.07 m thick layer of sediment may be reworked annually across the whole of the Dutch Wadden Sea at a density of 17 lugworms (Arenicola marina) m^{-2}, the average for this region (Cadée, 1976). The form

Table 7.2 Bioturbation functional groups after Bouchet et al. (2009).

Taxa	Description	Type	Ref.
Bivalves (*Macoma balthica; Mya arenaria*)	Move sediment in a random manner over short distances, resulting in diffusive transport	Biodiffuser	Michaud et al. (2005)
Polychaete worms (*Alitta virens*)	Dig tube system, resulting in non-local transport of matter from surface to deep parts of the tube	Gallery-diffuser	Michaud et al. (2005)
Sipunculida worms Tubificid oligochaete worms	Active transport of particles through the gut and passive transport as they move through the sediment	Downward conveyer Upward conveyer	Smith et al. (1986) Nogaro et al. (2006)
Crustaceans (e.g. Fiddler crabs)	Gallery-digging species that also cause biodiffusive mixing by releasing large amounts of sediment into the water column during digging	Regenerator	Gardner et al. (1987)

and structure of worm burrows are species-dependent, ranging from simple "J" and "U"-shaped burrows, to large galleries with multiple branches. For instance, *A. marina* constructs J-shaped burrows up to 0.2 m deep, ingesting sediment that slides down from the surface through a funnel (Volkenborn & Reise, 2006). Organic material is consumed and the resulting sediment is defecated to the surface as faecal mounds above the tail shaft of each burrow (Volkenborn & Reise, 2006). Infaunal bivalves move both vertically and horizontally through fine-grained sediments (Amyot & Downing, 1997). Many species feed on subsurface organic material, ejecting sediment into overlying layers. Many crustaceans are conspicuous bioturbators of sediment and can cover large areas with pseudofaecal pellets (Katz, 1980; Botto & Iribarne, 1999, 2000). For example, *Neohelice granulata* individuals process up to 5.9 kg of sediment m^{-2} day^{-1}, constructing funnel-shaped burrow openings 0.14 m in diameter and 0.4 m deep (Iribarne et al., 1997; Bortolus & Iribarne, 1999). Shrimp in the infraorders Gebiidea and Axiidea (formerly Thalassinidea) can occur in densities exceeding 200 m^{-2} (Branch & Pringle, 1987; D'Andrea & DeWitt, 2009) and construct large, complex burrows over 2 m deep, processing the equivalent of 12.14 kg m^{-2} d^{-1} (Branch & Pringle, 1987; Griffis & Suchanek, 1991; Dworschak & Ott, 1993; Nickells & Atkinson, 1995).

7.4.2   Controls on bioturbation by mobile benthic animals

Burrowing behaviour, and consequently bioturbation by mobile benthic animals, is not only species-dependent, but also varies with season, reproductive cycle, flow regime, substrate type, substrate disturbance, biological interactions and parasite abundance (Kat, 1982; Lewis & Riebel, 1984; Amyot & Downing, 1997; 1998; Di Maio & Corkum, 1997; Watters et al., 2001; Taskinen & Saarinen, 2006; Allen & Vaughn, 2009). Consequently, any impact on the physical environment will be controlled by the abiotic conditions and biotic interactions detailed below. These will need to be carefully considered in the design of experiments focused on the mixing of sediments by animals. It is also important to consider these factors when extrapolating rates and quantities of sediment disturbed by single or small groups of animals to entire populations over large areas. This is because the intensity of bioturbation is likely to be highly variable in time and space at a range of scales and between individual animals, both within and between species.

7.4.2.1   Size and density of organisms

In general, larger species have a greater impact on the environment than small ones (Dworschak & Pervesler, 1988; Solan *et al.*, 2004; Thrush *et al.*, 2006; Gilbert *et al.*, 2007; Bouchet *et al.*, 2009). Size variability within species is just as important as that between species. For example, large mussels move greater distances than small ones and mix sediments at faster rates (McCall *et al.*, 1995; Allen & Vaughn, 2009). Similarly, the size of burrows is dependent on the size of the individual. For instance, larger Thalassinidean shrimp (*Biffarius filholi* and *Callianassa subterranea*) dig larger burrows than smaller ones (Rowden *et al.*, 1998a; Berkenbusch & Rowden, 1999). Population density is also important for many bioturbators. For instance, Allen & Vaughn (2009) found that vertical movements by bivalves were greater at medium- and high-densities (19 and 39 mussels m^{-2}, respectively) compared with low densities (10 mussels m^{-2}). Consequently, in experiments, the size of individuals needs to be considered carefully. The median is a robust measure of central tendency in small populations or populations with an unknown distribution function, and so theoretically, the use of animals that are similar in size to the population median should provide the best estimates of average reworking. The areal density of animals (number of individuals m^{-2}) should ideally be equivalent to the density of organisms found under natural conditions, because this incorporates the important impact of biotic interactions, such as competition.

Water temperature exerts an important control on the burrowing of macrozoobenthos. Thalassinidean shrimp are less active in low water temperatures (<7°C), with implications for the scalability of annual sediment budgets from short-term laboratory experiments (Swinbanks & Luternauer, 1987; Rowden *et al.*, 1998b; Berkenbusch & Rowden, 1999). Above-optimal temperatures also decrease burrowing by shrimp and may explain the low burrowing activity of *Biffarius filholi* recorded in summer months by Berkenbusch & Rowden (1999). Temperature is partly responsible for the seasonal cycles observed in bioturbation by many macrofaunal species. Rowden *et al.* (1998a) calculated the rigidity modulus of the top 60 mm of field-collected sediment cores and found that the activity of brittle stars (*Amphiura filiformis*) was inversely correlated to bed rigidity. As a result, the rigidity modulus was 45% greater in January than in May, resulting in the bed being less resistant to erosion in summer and thus more likely to be disturbed during summer storms (Rowden *et al.*, 1998a). Wheatcroft *et al.* (1994) observed that the vertical distribution of tracers differed between cores recovered in spring and autumn, with those in spring decreasing monotonically with depth and autumn cores having a more uniform vertical concentration of tracers. To ensure that estimates of mixing potential are realistic, water temperature during experiments must be considered.

7.4.2.2   Sediment type

Sediment type and grain size are important controls on the impact of burrowing animals. All animals have limits to their strength and, consequently, they are only capable of reworking sediments up to a certain threshold grain-size and weight. For example, Wheatcroft (1992) found that glass beads with diameters of 8–16 μm were transported greater vertical distances by macrozoobenthos than coarser glass beads (126–420 μm). Johnson *et al.* (2010) found that the largest grain size signal crayfish (*Pacifastacus leniusculus*) could move from a uniform bed was 32 mm in diameter, with a submerged weight six times that of the average crayfish used in their experiments. Sediment cohesion is also an important influence on burrow structure and complexity because burrows in non-cohesive sediment will collapse without the addition of stabilising secretions or structures. The Thalassinidean shrimps *Neotrypaea californiensis* and *Neotrypaea gigas* both construct larger volume burrows in fine muds compared with those in coarser sands and *Pestarella candida* creates more complex burrows in fine sands than in coarse sands (Farrow, 1971; Griffis & Chavez, 1988). As a result of sediment preferences, distinct communities of bioturbating organisms can occur. An example is the survey by Hughes & Atkinson (1997) that found offshore muddy sands supported the shrimp *Callianassa subterranea* whereas the population shifted to *Upogebia deltaura* and the crab *Goneplax rhomboides* in coarser sediments farther inshore. Softer muds were characterised by mounds built by the echiuran *Maxmuelleria lankesteri* and burrows of the lobster *Nephrops norvegicus* (Hughes & Atkinson, 1997). Therefore, it is important to study the impacts of animals on their preferred host substrates as this will change their burrowing behaviour and the architecture of the resulting structures.

7.4.2.3   Location

Burrow characteristics differ with water depth. For instance, Dworschak (1987a) observed that burrows of *Upogebia pusilla* were deeper (0.8 m) in high intertidal areas than those in subtidal areas (0.2 m). Similar patterns have also been observed in *Callichirus major* and *Biffarius filholi* (Bradshaw, 1996; Berkenbusch & Rowden, 1999). This is likely to reflect the need to dig deeper in high intertidal areas to ensure moisture is maintained in burrows (Katrak *et al.*, 2008).

7.4.3   Quantifying bioturbation

Most studies of bioturbation have been undertaken in the field because deep sediments (>0.2 m) are necessary, but several studies have used still-water laboratory tanks and aquaria for this type of research. Several different techniques have been used to quantify the process. Resin casts of burrow systems, particularly those of crustaceans (Atkinson & Chapman, 1984; Meadows *et al.*, 1990; Davey, 1994; Astall *et al.*, 1997; Dworschak, 2002), have been used to quantify burrow architecture and volume. However, there are problems with ensuring resin penetration into deep burrow areas and with the blockage of smaller burrow shafts by air bubbles. The technique also destroys the burrow and surrounding sediment when the resin cast is removed from the bed.

Photographic techniques are less destructive. These include sediment profile imaging that allows visualisation of cross-sections through burrows and X-ray imaging to visualise intact burrows in sediment cores (Davey, 1994; Migeon *et al.*, 1999; Rosenberg & Ringdahl, 2005). However, these only give a 2D image of a 3D structure. Recently axial tomodensitometry (CT-scans and CAT-scans) has been used to visualise burrows in three-dimensions, allowing volumes and surface areas to be accurately ascertained and modelled (Perez *et al.*, 1999; Mermillod-Blondin *et al.*, 2003; Dufour *et al.*, 2005; Rosenberg *et al.*, 2007, 2008; Mazik *et al.*, 2008; Bouchet *et al.*, 2009). The detail available within the resultant images permits the study of small-scale burrows which are destroyed by other techniques. For instance, Mazik *et al.* (2008) were able to measure meiofaunal burrow volumes (0.1–1.3 mm^3) and measure burrow surface area (3.4–33.52 mm^2) using "microCT". However, the high cost of scans means that few cores can be sampled. The nature of the technique also means that cores must fit into the machine, limiting the size of burrow that can be scanned.

One of the most common methods for quantifying bioturbation uses tracers. In these studies, a bioturbation coefficient (D_b), defined as the rate at which the spread of particles in a tracer profile changes over time, is determined (Teal *et al.*, 2008). Radioisotopes, commonly including Pb^{210} and Th^{234} have been used to study bioturbation over timescales of weeks to centuries (Aller & Cochran, 1976; Benninger *et al.*, 1979; Kershaw *et al.*, 1984; Smith & Schafer, 1984; Rice, 1986; Wheatcroft *et al.*, 1994; Gouleau *et al.*, 2000). However, the records provided by this method integrate mixing over long time periods and so cannot be used to study short term processes, such as the lifespan of an individual animal. They also do not allow individual particles to be traced. Recently, tracer particles such as coloured sediments or glass beads have also been used. Similarly, luminophores are natural particles painted with fluorescent paint and are commonly used tracers in studies of bioturbation (Gerino, 1990; Quintana *et al.*, 2007). Other tracer methods include labelling individual sediment size fractions (e.g. silt, sand) with noble metals (e.g. Au and Ag), spreading the resultant tracers onto the sediment and recovering cores at a later date (Wheatcroft *et al.*, 1994).

Comparison of bioturbation between studies and species is often achieved with simple parameters, such as the burrow depth, horizontal extent and volume. However, the diversity of methods used, the variability in bioturbation due to abiotic and biotic controls, and the species-specific impacts of burrowing, make comparisons among studies difficult. For example, numerous researchers have calculated the rate at which sediment is processed by Thalassinidean shrimp (e.g. Swinbanks & Luternauer, 1987; Rowden *et al.*, 1998a, 1998b; Berkenbusch & Rowden, 1999). However, estimates of turnover rates use different methods and a variety of different units (Rowden & Jones, 1993). Understanding bioturbation at large spatial and temporal scales is challenging as studies are often short term and made over a limited area. For example, the amount of sediment expelled from shrimp burrows is high during construction but drops significantly once burrows are complete and then only increases for short periods during burrow maintenance (Stamhuis *et al.*, 1997). Consequently, the timing of estimates from short-term observations may exaggerate or underestimate sediment turnover rates.

7.5   TOPOGRAPHIC CHANGES DUE TO ANIMALS AND THE HYDRAULIC IMPLICATIONS OF THOSE CHANGES

7.5.1   Significance of animals in controlling substrate topography

Tubes, burrows, pits, tracks, mounds and faecal castings are created by animals at scales ranging from millimetres to metres and are ubiquitous in aquatic environments (McCall *et al.*, 1979; Rhoads, 1974; Richardson *et al.*, 1993; Willows *et al.*, 1998). These biogenic structures alter bed topography (surface roughness), with important implications for near-bed hydraulics and sediment stability (section 7.4). Aggregations of sessile animals, such as mussel and barnacle reefs, can alter surface roughness over hundreds of square metres (Meadows & Shand, 1989; Warwick *et al.*, 1997; Cummings *et al.*, 1998; Murray *et al.*, 2002). The significance of increased roughness for altering hydrodynamic conditions is dependent on the scale focus of the study. Increased roughness and turbulence generation by animals is important at the scale of the organism because it mixes overlying water layers, maintaining seston concentrations within the range of feeding appendages. However, the significance of roughness generation at larger scales is less well understood. For instance, most laboratory studies have focused on the impact of mussels in flows that are less than 1 m deep and, therefore, their significance in deeper marine environments has not been established.

Roughness generation by the constructions of mobile animals is less well studied, when compared with sessile animals because mobile animals create comparatively small local impacts (Table 7.3). However, mobility means that the impacts occur over an area far larger than their body surface. Roughness generation by mobile animals may thus be viewed as an extensive process, while for sessile animals it may be viewed as an intensive process. In the absence of quantitative information on the hydraulic environment over and biogenic roughness elements, height can be used as a measure of the potential impact on the hydrodynamics of the system. Despite their comparatively small scale, protruding faecal pellets and biogenic constructions generate turbulence near the bed, with implications for bed shear stress and the hydrodynamic environment (Nowell & Jumars, 1984). Rowden *et al.* (1998b) used photography to estimate a boundary roughness length (z_0) of 7.9 mm over mound fields of Thalassinidean shrimp, in contrast to only 0.007 mm over plane beds. Sessile animals generate roughness with their bodies and shells and can increase z_0 by many millimetres compared with bare sediment. For example, Peine *et al.* (2011) reported that roughness length (z_0) was increased by a factor of 10^4 over *Mytilus edulis* aggregations. However, the presence of biogenic topography can suppress the development of bedforms. For example, the removal of fauna from Dry Tortugas Bay (Florida Keys, USA) led to an increase in z_0 because ripples developed on the sediment surface (Wright *et al.*, 1997).

7.5.2   Ways in which animals alter substrate topography

The nature and impact of topographic alterations by animals is dependent on the species involved. For instance, polychaete worms create tubes that project from sediment surfaces whereas some burrowing bivalves create pits and mounds. However,

Table 7.3 Roughness height and roughness length over infaunal (mobile) and epifaunal (sessile) bivalve species. Roughness height is the difference between the highest and lowest point. Roughness length (z_0) is the height above the bed where flow velocity equals zero. This can be estimated by extrapolating from a logarithmic velocity profile or using Nikuradse's parameterization $z_0 = D_{50}\ 30^{-1}$, where D_{50} is the median grain size. Results calculated with the latter method are marked with*.

Species	Type	Density (m^{-2})	Roughness height (mm)	Roughness length (mm)	Reference
Austrovenus stutchburyi	Infaunal clam	1000	10		Jones et al. (2011)
Cerastoderma edule	Infaunal cockle	312	20		Cuitat et al. (2007)
		300		1.5	Fernandes et al. (2007)
Crassostrea gigas	Epifaunal Oyster		100–200		Troost et al. (2009)
Mytilus edulis	Epifaunal mussel	755		1*	Butman et al. (1994)
		1800	25–30	0.8–6.9	Van Duren et al. (2006)
		177		5–40*	Peine et al. (2011)
Atrina zelandica	Epifaunal mussel	100		4–22	Green et al. (1998)
		85	120		Nikora et al. (2002a)
		12.5	60		

the great variety of topographic features generated by animals can be broadly divided into the four categories shown in Table 7.4 and discussed below.

7.5.2.1   *Aggregation of sessile animals*

Mussel reefs and beds have three major components: a topographically complex matrix of dead and living shells that are connected by a mat of byssal threads and can attain considerable thickness; an underlying layer of sediment, faeces and pseudofaeces; and a surface layer of associated organisms (Figure 7.4; Suchanek, 1985, 1992; Lintas & Seed, 1994; Seed, 1996). The structural stability of reefs, coupled with topographic complexity, sheltering and increased hydraulic heterogeneity also make them important autogenic ecosystem engineers, often supporting much expanded communities of organisms relative to surrounding areas (Kidwell, 1986; Crooks & Khim, 1999; Gutiérrez & Iribarne, 1999; Lenihan, 1999; Gutiérrez *et al.*, 2003; Norkko *et al.*, 2006). Some of these associated organisms, such as barnacles (*Semibalanus balanoides*) and bivalves (*Lasaea adansoni*) (Lintas & Seed, 1994; Cummings *et al.*, 1998) may add considerably to the roughness and topographic complexity of a reef (Figure 7.4).

Table 7.4 Common biogenic structures constructed by mobile animals in benthic environments (images not to scale).

	Animal reefs	*Tracks and pellets*	*Pits and mounds*	*Polychaete worm tubes*
Image				
Example taxa	Mussels, barnacles	Fiddler crabs	Thalassinidean shrimp	Polychaete worms
Roughness length (mm)	0.8–6.9		0.79	0.023–0.85
Diameter (mm)		3–12	300	2
Vertical Extent (mm)	30 mm	3–12 diameter	−300 to +500	12.5
Population density (structures m⁻²)	1800	40–300 burrows		1000 tubes
Reference	Van Duren et al. (2006)	McCraith et al. (2003)	Rowden et al. (1998b)	Peine et al. (2009)

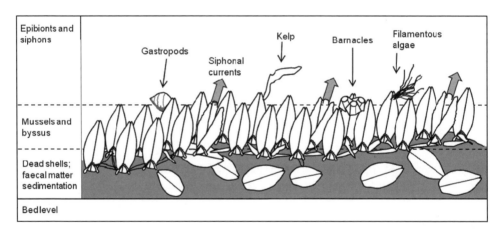

Figure 7.4 Diagram of the three main components of a mussel bed: 1) An underlying area associated with sedimentation and the build-up of faecal and pseudofaecal material; 2) An overlying layer of live and dead mussels connected with a mat of byssus threads; and 3) A final layer of attached and associated organisms, potentially increasing roughness.

The increased height of bivalve reefs above the bed accelerates flow, with increases of up to 64% over an artificial oyster reef reported by Lenihan (1999). The increased roughness associated with bivalve beds also increases the thickness of the boundary layer (Nikora *et al.*, 2002a). Nikora *et al.* (2002a) reported that the flow in the boundary layer above mussels is characterised by significantly reduced longitudinal velocities and greater turbulence intensities in comparison to boundary layers in areas without mussels. Turbulence is generated at the abrupt roughness changes along the boundary between sessile animal reefs and bare sediment (Folkard & Gascoigne, 2009).

Butman *et al.* (1994) reported that the turbulence intensity, quantified as the root mean square variability about the time- (or Reynolds-) averaged velocity, was three times greater in slow flows (5 cm s^{-1}) and ten times greater in fast flows (0.15 m s^{-1}) above mussel beds than over the smooth flume bottom. Widdows *et al.* (2009) found that mussel patches in the field significantly increased bed roughness, with resultant increases in TKE and bed shear stress. Turbulence generation at patch boundaries has implications for the concentration of organic matter at the bed, with increased turbulent mixing potentially preventing depletion of seston in the boundary layer (Widdows *et al.*, 2009).

Sparsely distributed protruding objects have an impact on the flow that is equivalent to the sum of individual flow disturbances (Nowell & Jumars, 1984). However, when objects of similar height occur in a sufficient spatial density, the wake and associated eddies shed from the objects interact, producing a change that is greater than the sum of individual objects (Crimaldi *et al.*, 2002; van Duren *et al.*, 2006; Folkard & Gascoigne, 2009). If the density is further increased then the flow will skim over troughs in the roughness structure rather than through them, effectively reducing roughness, known as skimming flow (Figure 7.5; Nowell & Church, 1979; Eckman, 1983; Nowell & Jumars, 1984; Friedrichs *et al.*, 2000; Coco *et al.*, 2006). Green *et al.* (1998) found that the drag coefficient was nearly four times larger over horse mussel (*Atrina zelandica*) beds than flat, abiotic beds, but was lower over beds with the highest mussel density due to the development of skimming flow. This is supported by research into the fractal characteristics of mussel topography, which indicates that the total surface roughness (characterised by the fractal dimension) is higher for patchy distributions of *Mytilus edulis* at intermediate densities than for a

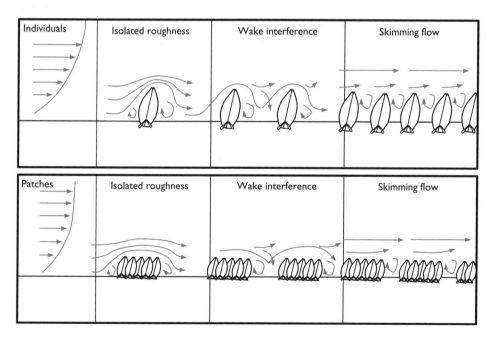

Figure 7.5 The impact of reducing the spacing between individuals and patches of mussels on surface roughness and near-bed flow.

complete cover of similar-sized, densely packed mussels (Snover & Commito, 1998; Commito & Rusignuolo, 2000; Crawford et al., 2006; Figure 7.5). Because skimming flow can impair filter-feeding by limiting turbulent mixing of near-bed flows, which reduces the supply of seston, many sessile organisms develop patterns that disrupt it. Barnacles form hummocks at high population densities that act as bluff bodies, generating a heterogeneous, highly turbulent flow environment that increases seston transport across the whole colony (Pullen & LaBarbera, 1991; Thomason et al., 1998). Similarly, mussel beds can form complex, irregular patterns of clumps and strings (Paine & Levin, 1981; Wootton, 2001; van de Koppel et al., 2005), representing self-organised patchiness (Gascoigne et al., 2005; van de Koppel et al., 2005; 2008; van Wesenbeeck et al., 2008). Therefore, it is important in experiments with live sessile organisms to allow them to acclimatise to experimental conditions in order for them to attain a spatial organisation equivalent to that of a field setting.

7.5.2.2 Animal tracks and surface pellets

At low population densities (29 to 100 individuals m^{-2}), the tracks generated by an infaunal, suspension-feeding, clam (*Austrovenus stutchburyi*) found on intertidal sand and mudflats in New Zealand generated bed roughness of the order of 10 mm (Jones et al., 2011). At higher population densities (747 to 1000 m^{-2}), roughness increased because of partially-exposed mussels (Jones et al., 2011). This increased the boundary layer height, reduced flow speeds and increased bed shear stress at flow velocities of 50 and 150 mm s^{-1} but did not have an impact at 20 mm s^{-1} (Jones et al., 2011). The presence of faecal pellets and coils on sediment surfaces increases surface roughness and Nowell et al. (1981) observed pellets that disrupted the viscous sublayer. Photographic surveys of the seabed indicate that animal tracks and pellets are ubiquitous and important components of substrate topography, particularly in abyssal regions where physical disturbance is otherwise minimal (Smith et al., 1997; Przeslawski et al., 2012).

7.5.2.3 Pits and mounds

Many animals, including crustaceans, worms and bivalves, create pits and mounds on the surface of sediment beds. For instance, the bivalve *Macoma balthica* creates feeding depressions approximately 2 mm deep (Widdows et al., 1998a) and lugworms (*Arenicola marina*) construct pit and mound topography in sandy sediments (Volkenborn & Reise, 2006). Although the flow through these structures and their impact on the near-bed hydraulic environment and sediment stability has not been suitably quantified, studies of the flow around other objects that protrude into the flow, for example pebble clusters in rivers (Brayshaw et al., 1983; Buffin-Belanger & Roy, 1998), may provide surrogate examples of the potential impact of biogenic mounds. In particular, it is likely that these features increase the heterogeneity of near-bed flows, creating local areas of flow acceleration and deceleration and generating flow structures. Pits and other biogenic depressions are likely to reduce velocities due to the locally increased depth and, depending on the depth of the depression, create areas of recirculation. This will enhance deposition and trap transported sediment, both of which may provide organic material to organisms (Yager et al., 1993).

The altered topography also creates pressure differences that can passively irrigate burrow systems (section 7.3; Ziebis *et al.*, 1996). In addition, Huettel & Gust (1992), using dye tracers, found that biogenic roughness increased porewater fluxes in sandy sediments by generating small-scale areas of upwelling and downwelling. Such features are extremely important for solute exchange, nutrient cycling and oxygenation of ocean sediments (Figure 7.6).

Thalassinidean shrimp occur in dense colonies and mix large quantities of sediment by grazing sediments (Dworschak, 1987b) which are deposited at the burrow opening, forming large conical mounds. Over 47% of 44 species reviewed by Griffis & Suchanek (1991) deposited mounds at burrow openings, with those of *Neocallichirus maryae* reaching 0.19 m in height at a spatial density of 6 to 7 mounds m⁻² (Suchanek, 1983; Suchanek *et al.*, 1986). Some burrows also have funnel-shaped openings, for instance, *Glypturus acanthochirus* forms craters 0.5 m in diameter and 0.3 m deep that occur at densities of 5 m⁻² (Dworschak & Ott, 1993). Pits and mounds are also constructed in freshwater systems. For example, crayfish can burrow into fluvial substrates and riverbanks, with implications for bed and bank stability (Guan, 1994; Statzner *et al.*, 2000, 2003; Harvey *et al.*, 2011). Still-water experiments in aquaria and flume studies in flowing water found that signal crayfish excavate pits in gravel substrates (diameters 8–22 mm) and heap excavated material into mounds and ridges (Figure 7.7; Johnson *et al.*, 2010; 2011).

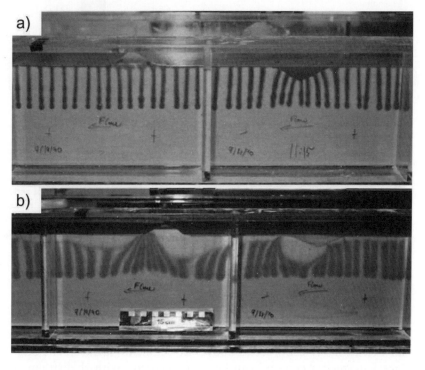

Figure 7.6 Dye patterns produced by advective porewater flows associated with a mound (20 mm high) and a funnel (20 mm deep) in sandy sediment. a) Dye pattern at the start of the experiment and b) after 3 hours of flow. Flow direction right to left (from Huettel & Gust, 1992).

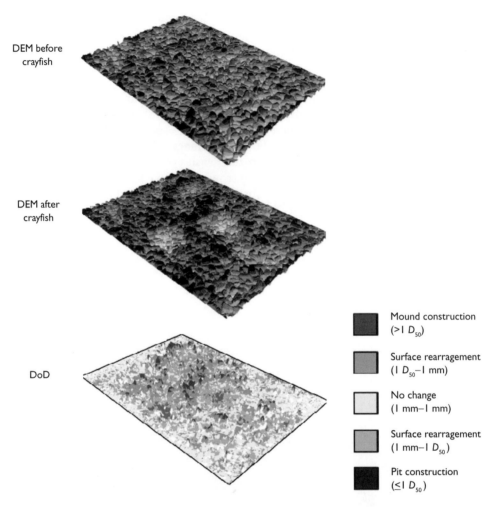

Mound construction
($>1 D_{50}$)

Surface rearragement
($1 D_{50}$–1 mm)

No change
(1 mm–1 mm)

Surface rearragement
(1 mm–$1 D_{50}$)

Pit construction
($\leq 1 D_{50}$)

Figure 7.7 DEMs of 11–16 mm gravel surfaces (0.4 × 0.6 m) before and after 6 hours of activity by a single signal crayfish (Pacifastacus leniusculus) in a low velocity flow (0.1 m s⁻¹). The resultant DEM of Difference (DoD), determined by subtracting the DEM before crayfish from the DEM after crayfish, is shown below and shaded according to the magnitude of topographic change.

7.5.2.4 Tubes

Many species of worm, particularly polychaete worms, create tubes that can generate significant bed roughness. Morphology depends on worm species, but tubes can extend many centimetres above the sediment surface. The area above the surface is termed a cap and can be "ornamented" with shell fragments and algae (Bell, 1985). Polychaete worms can occur in high densities, creating "lawns" of tubes (Fauchald & Jumars, 1979). The altered hydrodynamic environment increases fluxes of dissolved and particulate matter at the sediment-flow interface, with implications for feeding (Huettel & Gust, 1992; Huettel *et al.*, 1996; Huettel & Rusch, 2000). Some species construct rigid tubes that stand erect in flows whilst others construct flexible ones (Fager, 1964; Eckman *et al.*, 1981). This may result in some tube lawns inducing a canopy layer comparable to those observed in seagrass (Chapter 6). Some characteristics of worm tubes are summarised in Table 7.5.

Scaled flume results of Peine *et al.* (2009) suggest that the presence of biogenic structures, including worm tubes, increase the roughness length by a factor of 2 to 30 in the Baltic Sea. Tubes can have complex impacts on the hydrodynamic environment, with implications for bed stability, morphology and sedimentation (section 7.6). Flume experiments have identified that flow accelerates horizontally around an individual tube and that horseshoe vortices develop, deflecting high momentum fluid towards the bed and scouring non-cohesive sediments (Eckman *et al.*, 1981; Eckman & Nowell, 1984). The height of the tube affects the magnitude of the disturbance. The RD of tubes, defined as the ratio of tube area to sampling area and expressed as a percentage, has a controlling influence on their hydrodynamic impact. Peine *et al.* (2009) found that RDs of surrogate tubes of 0.7–1.9% had substantial impacts on the hydrodynamic environment, increasing turbulence generation due to interference between individual roughness elements. As RD increased from 4.2 to 7.5%, skimming flow began to develop, with a reduction in turbulence generation and protection of underlying sediments (section 7.6). Worm tubes that are clustered together can form hemispherical mounds, similar to those found in dense aggregations of barnacles and mussels, which modify the hydrodynamic environment increasing feeding currents in comparison to isolated individuals and limiting the potential for competition (Merz, 1984).

Table 7.5 Worm tube dimensions and densities (not including surrogates).

Species	Density (m⁻²)	Flexibility	Diameter (mm)	Length (mm)	Reference
Melinna cristata	5000	Rigid			Buchanan (1963); Fauchald & Jumars (1979)
Lanice conchilega		Rigid			Carey (1983)
Spio setosa	2000	Rigid			Muschenheim (1987)
Polydora ciliate		Flexible	2	10	Hempel (1957); Peine *et al.* (2009)
Lagis koreni			2	7.5	Peine *et al.* (2009)
Owenia fusiformis	500–1000	Flexible	0.6–5		Fager (1964); Eckman *et al.* (1981)
Pygospio elegans	1000		0.75–2	12.5	Peine *et al.* (2009)
Diopatra cuprea		Rigid		10–60	Myers (1972); Luckenbach (1986)
Diopatra leuckarti	21800	Rigid		10–20	Bailey-Brock (1984)
Polychaete spp.	440–3670	Rigid	2.5	10	Jørgensen *et al.* (2005)

7.5.3   Quantifying topographic alterations by mobile animals

Roughness has been quantified in two main ways, through calculation of the "roughness height" and "roughness length". Roughness height is the distance between the highest and lowest points on the roughness element. Roughness length (z_0) is calculated by measuring velocity at a series of heights through the water column and then fitting the logarithmic law of the wall to the resulting profile. By extrapolating from this profile, the intercept along the y-axis indicates the height above the bed where velocity is zero. This is only an estimate of z_0, but it is regularly used and is semi-quantitative. An alternative method for estimating z_0 employs Nikuradse's parameterisation:

$$z_0 = D_{50}\ 30^{-1} \tag{7.1}$$

where D_{50} is the median size of the roughness element. This measure is entirely dependent on the size of the roughness element and, therefore, ignores feedback between the flow field and the size of roughness elements in controlling the size of the boundary layer. A novel way of quantifying surface roughness is using fractal analysis. The fractal dimension is a measure of surface complexity and organisation and has been applied to mussel beds to quantify roughness at a range of densities (Snover & Commito, 1998; Commito & Rusignuolo, 2000; Crawford et al., 2006).

Roughness has been assessed in both field and flume studies, but the detailed hydrodynamic impact of increased roughness has mostly been studied in flumes. An important exception is the work of Nikora et al. (2002a), who studied horse mussel (Atrina zelandica) beds in Mahurangi Harbour, North Island, New Zealand using detailed, spatially-distributed ADV measurements.

Briggs et al. (2001) used stereophotos and Johnson et al. (2010; 2011) used a laser scanner to obtain 3D images of surface topography. The latter used the high resolution 3D topographic datasets to create Digital Elevation Models (DEMs) of sediment surfaces before and after the activity of signal crayfish (Pacifastacus leniusculus) (Figure 7.7). However, laser scans cannot routinely be obtained through water (although see Smith et al., 2012), limiting their use to either sediments that can be removed from field environments or laboratory experiments. Draining a hydraulic facility to obtain a laser scan or make other bead measurements requires care because the dewatering of sediments may alter their structure and composition.

In all studies of animal-generated roughness, the environmental context of the animal is essential to ensure results are representative of field environments. In laboratory experiments, van de Koppel et al. (2008) found that mussels (Mytilus edulis) actively moved into small aggregations. Once these small clusters had developed, movement decreased and mussels became interconnected with byssal threads, hypothesised to be a mitigating tactic against flow forces. However, as population density continued to increase and clusters became larger, mussel movement increased again, reducing aggregations, which was attributed to food depletion. Intermediate population densities of mussels in patchy aggregations may also provide better protection from dislodgement. For instance, painted tracer mussels released into dense, homogenous beds were lost due to flow or wave action whereas those released onto patterned beds persisted for significantly longer periods of time (van de Koppel et al., 2005). Consequently, acclimatising animals to experimental conditions, and using distributions and densities of organisms similar to those found in field settings, are essential to obtain results analogous to field equivalents.

7.6   THE IMPACT OF MOBILE ANIMALS ON THE STABILITY AND TRANSPORT OF SEDIMENT

7.6.1   The significance of animals for the stability of freshwater and marine sediments

Communities of animals and individual species are highly significant in controlling sediment transport in marine systems. In many situations, particularly in fine-grained, low-energy environments, this may be as important a control as diurnal or lunar tidal cycles (Swift, 1993; Widdows *et al.*, 2000a, 2000b). For instance, DeBacker *et al.* (2010) found that sediment stability in Ijzermonding tidal flat, Belgium, varied over short (<1 year) timescales and the macrofauna, dominated by mud shrimp (*Corophium volutator*), and algal biofilms were most important for stability. Macrofauna alter stability through direct disturbance associated with burrowing, the flushing of burrow sediments and the grazing of stabilising microbial biofilms (DeBacker *et al.*, 2010). However, disentangling the direct impact of macrofauna and their often complex relationships with meiofauna and algal communities is challenging.

7.6.2   How animals alter sediment stability and sediment transport

Alterations to surface topography, the near-bed hydrodynamic environment and sediment composition and structure can alter the stability of the substrate. Vertical and horizontal movements of benthic animals and the pelletisation of surface sediments can break up and loosen cohesive sediments, creating an erodible surface layer in marine environments termed a "fluff layer" (Willows *et al.*, 1998; Orvain *et al.*, 2003b; Orvain, 2005). The impact of animals near-bed hydrodynamics also indirectly causes material to be removed from the water column, by inducing sedimentation. In addition, animals consume sediments and organic matter, often capturing material from suspension, and transform and deposit it as faecal and pseudofaecal material. This is termed biodeposition and is important in the biogeochemistry of aquatic environments and responsible for the transformation of large quantities of organic and inorganic material. Animals interacting with non-cohesive sediments can destabilise them by altering sand and gravel matrix structure, topography, rigidity and water content (Statzner *et al.*, 2003; Johnson *et al.*, 2011). In some cases they may also favour the stabilisation of sediment, for instance, by reducing turbulence generation at the bed or by armouring sediments. Finally, many animals secrete sticky substances that bind grains together and stabilise the bed (Wotton, 2011).

7.6.2.1   *Suspension of fluff layers*

The instability of fluff layers results in their redistribution by waves and currents during almost every tidal cycle, with surface layers behaving more like a suspension of mucous-rich flocs rolling over the bed (Denny, 1983). Fluff layers are caused by the burrowing and foraging activity of numerous marine taxa and the addition of faecal and pseudofaecal material that can make up 80% of these layers (Rhoads & Young, 1970a; 1970b; Davis, 1993; Blanchard *et al.*, 1997; Andersen, 2001a; Orvain *et al.*, 2003b; Orvain, 2005). For example, the common mud snail (*Peringia ulvae*), a common deposit feeder across European intertidal mudflats, alters surface roughness and stability by creating track-ways and faecal mounds on the sediment surface (Nowell, 1981; Blanchard *et al.*, 1997; Andersen, 2001a; 2001b; Andersen *et al.*, 2002; Orvain *et al.*, 2003b; 2004; 2006). Andersen *et al.* (2002) found that population densities of 10000 and 50000 m^{-2} lowered the threshold for entrainment and increased the erosion rate of cohesive sediment beds by a factor of 2 to 4.

Many organisms directly eject fine-grained sediments into the water column, enhancing sediment resuspension and water turbidity and potentially smothering other organisms (Colin *et al.*, 1986; De Deckere *et al.*, 2000). Duarte *et al.* (1997) calculated that *Callianassa* shrimp reworked the substrate within an entire seagrass meadow in two years. Shrimp store coarse material in deeper burrow areas and preferentially eject finer sediment to the surface where it can be more easily entrained (Siebert & Branch, 2006). This can result in a significant contribution to transport, calculated to be 11 kg (dry) m^{-2} year^{-1} for *Callianassa subterranea* in the North Sea (Rowden & Jones, 1994). Significantly greater suspended sediment concentrations have been recorded at sites with *Callianassa* in comparison to sites without (Suchanek & Colin, 1986; Riddle, 1988; Ziebis *et al.*, 1996; Rowden *et al.*, 1998b; Siebert & Branch, 2006). For instance, Murphy (1985) reported that *Neotrypaea californiensis* caused a 10-fold increase in suspended sediment concentrations. The amphipod *Corophium volutator*, which can occur in population densities of 100000 m^{-2}, not only significantly increases the resuspension of bed material, but was found to be more important than resuspension by the flow in velocities <0.2 m s^{-1} or when biofilms were present (De Deckere *et al.*, 2000).

7.6.2.2   Sedimentation and biodeposition

Deposition of fine material from the water column is induced over the surface of aggregations of animals due to the increased surface roughness, slower flows, and increased downward turbulent mixing. Commito *et al.* (2005) collected nearly three times more sediment from sediment traps within mussel beds than from traps outside of the beds. The small Asian species of mussel, *Arcuatula senhousia*, is now a widespread invasive species on intertidal and estuarine soft sediments in Australasia, the Mediterranean and the west coast of North America. It uses byssal threads to create bags and cocoons that can form a thick mat at high population densities (5000 to 10000 m^{-2}) and can alter bed sediment composition (Morton, 1974; Creese *et al.*, 1997; Crooks, 1996; 1998; Crooks & Khim, 1999). The accumulation of sediments within these mats can elevate them by as much as 0.1 m above the surrounding substrate (Creese *et al.*, 1997).

Aggregations of filter feeders, such as barnacles and bivalves, also produce and deposit large quantities of faecal material. Many filter-feeders also produce pseudofaeces, which consist of inorganic matter that has been filtered from the flow but subsequently ejected. For example typical oyster (*Crassostrea virginica*) beds can produce 981 kg of faeces and pseudofaeces every week with deposits containing 77–91%

inorganic matter (Haven & Morales-Alamo, 1966). In the (Dutch) Wadden Sea, *Cerastoderma edule* was found to deposit 100000 tonnes (dry weight) yr^{-1} (Verwey, 1952). ten Brinke *et al.* (1995) showed that mussels (*Mytilus edulis*) deposit 50–100 mm of fine-grained sediment in a summer, approximately equal to that deposited directly from the water column. The deposit-feeding bivalve *Yoldia limatula* expels thick, watery slurry several centimetres into the water column; an individual of median shell length can resuspend sediment with a total dry weight of 0.44 kg yr^{-1} (Bender & Davis, 1984). Biodeposition is therefore a highly significant process in marine systems, even at the largest scales. The selective filtering of coarse material and biodeposition of finer grains can result in alterations to sediment grain size composition, with implications for sediment stability. Widdows *et al.* (2009) found that 95% cover of *Mytilus edulis* reduced the disaggregated D_{50} of the substrate from 138 μm to 19 μm as a result of an increase in silt/mud content from 22% to 65%. Similar increases in silt content associated with combinations of biodeposition and sedimentation have been observed in other studies (Ragnarsson & Raffaelli, 1999; Ysebaert *et al.*, 2009).

7.6.2.3 Bulk sediment destabilisation

The shells of sessile animals, such as bivalves, are an important component of the bedload that can abrade substrate sediments. Bed erosion rates of cohesive substrates increased up to 20-fold with the introduction of a single transportable shell to an annular flume (Amos *et al.*, 2000). There was a linear increase in erosion rate with increasing shell size and an exponential increase in the suspended sediment concentration with time (Thompson & Amos, 2002). Sediment is also preferentially scoured from around immobile shells. For instance, the mean velocity required to entrain unconsolidated sand reduced by 40% in the presence of solitary surrogate shells (Friedrichs *et al.*, 2009).

Bioturbation can change the protrusion and orientation of sediment particles and hence the stability of the surface (Table 7.6). The presence of clams (*Euterebra tantilla*) for 15 hours reduced the velocity required to entrain cohesive sediments by 20% due to the production of 2 mm deep tracks that covered less than 10% of the surface area of the bed (Nowell *et al.*, 1981). Seasonal increases in *Macoma balthica*

Table 7.6 Impact of bivalves on sediment stability compared to beds without bivalves.

Species	Population density (m^{-2})	Sediment transport compared to plane bed	Ref.
Venerupis philippinarum	0–71	1700% increase	Sgro et al. (2005)
Cerastoderma edule	312	6000% increase at 30 cm s^{-1}	Cuitat et al. (2007)
Cerastoderma edule	25–100% surface cover	25% decrease in bed shear stress	Widdows et al. (2002)
Macoma balthica	1000	400% increase in resuspension	Willows et al. (1998)
Macomona liliana		200% decrease in stability	Lelieveld et al. (2004)
Scrobicularia plana		67% decrease in critical shear stress	Orvain (2005)

density, coupled with changing microphytobenthos densities, resulted in a 30- to 50-fold decrease in sediment stability in the Westerschelde Estuary, The Netherlands (Widdows *et al.*, 2004). At population densities of 1000 m⁻², similar to recorded densities on Skeffling mudflat in the Humber estuary, Willows *et al.* (1998) found that *Macoma. balthica* increased sediment transport by 400% at low current velocities (0.15 m s⁻¹) but only by 200% at higher velocities (0.35 m s⁻¹). Consequently, the significance of animals is likely to be controlled by flow rate, with an upper limit above which the activity of animals is no longer significant. Other abiotic parameters are also relevant. For instance, Blanchard *et al.* (1997) found that total sediment cohesion played an important role in controlling the impact of the snail *Peringia ulvae*; Orvain *et al.* (2003b) found that volumetric moisture content of sediments controlled sediment destabilisation by *Peringia ulvae*.

The invasive signal crayfish (*Pacifastacus leniusculus*) has been found to alter the stability of river gravels (8–16 mm) (Johnson *et al.*, 2011). Non-cohesive substrates become structured through the continuous action of fluid flow moving vulnerable grains into less vulnerable positions, which stabilises the bed by generating an imbricated and relatively coarse surface layer. Crayfish activity was found to reverse this process resulting in the mobilisation of nearly twice as many grains from crayfish-disturbed substrates than from controls without crayfish under a given flow (Johnson *et al.*, 2011).

7.6.2.4   Stabilisation of sediments

Many animals secrete or produce substances that can adhere to or bind sediments together (Wotton, 2011; Table 7.7). For instance, chironomid larvae build tubes from sand grains (Brennan *et al.*, 1979; Pringle, 1985) and polychaete worms produce sticky mucous and cement that bind grains and fragments of shell together (Fager, 1964). The crab *Neohelice granulata* traps clay and silt particles at its burrow opening, which increases sediment surface cohesion and stabilises sediments (Botto & Iribarne, 2000). Many insect larvae produce silk threads. In freshwater environments, caddisfly larvae produce large quantities of silk. Some caddisfly build cases that they carry around as portable shelters (Wiggins, 2004) whereas others, such as hydropsychid caddisfly, use silk to build safety lines, filter-nets, retreats and, eventually, cases within which larvae pupate (Statzner *et al.*, 1999; Cardinale *et al.*, 2004). Johnson *et al.* (2009) placed trays of gravel in the River Soar, UK for 21 days to allow them to be colonised by caddisfly larvae and then carefully removed and transported them to a laboratory flume. The entrainment of material from trays with caddisfly silk was then compared to controls. Silk threads bound fluvial grains together, increasing the shear stress required to entrain 4–6 mm gravels by 38% (Johnson *et al.*, 2009). This was significantly greater than that caused by abiotic stabilising processes, such as the deposition of fine sediment and the rearrangement of surface grains (Johnson *et al.*, 2009).

7.6.2.5   Bioprotection of sediments

Molluscs are ubiquitous in most marine and freshwater habitats (Dillon, 2000) and, as a result, mollusc shells are also extremely widespread (Vaughn & Hakenkamp, 2001; Gutiérrez *et al.*, 2003). Molluscs produce large amounts of calcium carbonate

Table 7.7 Mobile macrofauna that produce adhesive material including silk, mucous and/or byssus.

Taxa	Adhesive material	What	Effect	Ref.
Thalassinidean shrimp				
e.g. *Pestarella candida,* *P. Tyrrhena*	Burrow wall lining	Mucous secretion	Aggregate fine grains (<2 mm)?	Dworschak (1998)
Insect larvae				
e.g. *Hydropsyche* spp.	Filter nets, retreats, safety lines	Silk	Silk binds grains <6 mm	Johnson et al. (2009)
Polychaete worms				
e.g. *Hediste diversicolor*	Burrow wall lining	Mucous secretion	Binds grains together	Meadows et al. (1990)
e.g. *Phragmatopoma californica*	Burrow wall lining	Calcium and magnesium cement	Bonds grains together	Stewart et al. (2006)
Crustaceans				
e.g. *Corophium volutator*	Burrow wall lining	Mucous secretion	bind grains	Meadows et al. (1990)

when making and maintaining shells. For example, the cockle (*Cerastoderma edule*) produces 0.05–1 kg CaCO$_3$ m^{-2} yr^{-1}, and freshwater molluscs have also been reported to produce shell in similar quantities (Beukema, 1982; McMahon & Bogan, 2001). The shell production rate of some marine oysters is estimated to reach 9 kg m^{-2} yr^{-1} (Powell *et al.*, 1989). The durability and ubiquity of shells means they are pervasive in aquatic environments. For example, beds of stout razor clam shells (*Tagelus plebeius*), radiocarbon dated at 3850 ± 60 and 1340 ± 50 ya, have been uncovered by erosion of fine sediments in SW Atlantic estuaries in Argentina (Iribarne & Botto, 1998; Gutiérrez & Iribarne, 1999). Experimental removal of these shells led to substantial sediment destabilisation and increased erosion as well as a decrease in silt-clay content due to preferential erosion of finer sediment fractions (Gutiérrez & Iribarne, 1999). In addition, shells can be used to protect coastlines from erosion. For example, the addition of oyster (*Crassostrea virginica*) shells to lower fringes of marshes in North Carolina, USA, protected them from wave forces and stabilised sediments, resulting in a reduction in marsh loss (Meyer *et al.*, 1997). O'Donnell (2008) found that, relative to unsheltered conditions, wave-induced forces were reduced by 30–62% within a 100 mm diameter bare patch surrounded by tightly-packed aggregations of mussel shells. As the diameter of the bare patch was increased, this buffering effect was reduced and was not noticeable at a diameter of 300 mm (O'Donnell, 2008). Therefore, shells can form biogenic armour layers, protecting underlying sediments and coastlines.

Bivalves can be extremely securely attached to underlying substrates (Bromley & Heidberg, 2006). Many, including the blue mussel (*Mytilus edulis*), secrete byssus threads (Waite, 2002; Waite *et al.*, 2002). The tensile strength of these threads, which has been measured as 21 MPa (Price, 1981), allows bivalves to colonise areas associated with high turbulent energy (Witman & Suchanek, 1984). Many bivalves, including oysters, can also fix themselves to substrates with cement that is similar in composition to that of the shell (Yonge, 1960; Lehmann & Wippich, 1995; Yamaguchi, 1998). The stability of aggregations of bivalves, in combination with shell durability, essentially armours the sediments that they colonise (Wildish & Kristmanson, 1997; Widdows *et al.*, 1998b; 2002; van Leeuwen *et al.*, 2010). For example, the Asian clam *Arcuatula senhousia* produces byssus mats that can increase the shear strength of sediment by 70% in comparison to mat-free areas (Crooks, 1998).

Some species of polychaete worm, particularly the family *Sabellariidae*, create worm-reefs that are widespread globally (Fager, 1964; Rees, 1976; Posey *et al.*, 1984; Noernberg *et al.*, 2010). Reefs are constructed on firm ground such as rock, shells or firm clay and are composed of numerous tubes of cemented sand grains. The need for sand grains to construct reefs restricts them to intertidal or near-subtidal areas with sufficient wave energy to suspend sand (Kirtley & Tanner, 1968). Individual tubes are around 120 mm long and 1.7 mm wide and occur in densities of 750,000 m^{-2}, forming large, honeycomb-like, wave-resistant reefs that can reach 0.4 m above the substrate surface (Rees, 1976). As a result of these habitat alterations, *Sabellariidae* may be considered as ecosystem engineers with reefs harbouring a large range of other organisms, including urchins, nudibranches, polychaetes, crustaceans and gastropods (Posey *et al.*, 1984; Dubois *et al.*, 2006). The potential protection of underlying sediments and the reduction in wave-forces for residents living on these reefs is currently unknown, but is likely to be highly significant.

7.6.3   Quantifying the effects of animals on sediment stability and transport

Both field and laboratory studies have been undertaken to assess the significance of animals for the morphology, structure and stability of bed material. Hydraulic experiments often quantify the erosion of sediments as a mass per unit area (kg m^{-2}) or measure a sediment erosion rate (kg m^{-2} s^{-1}), usually in response to stepped increases in flow velocity (Widdows *et al.*, 2000a; 2000b). Most laboratory studies allow animals to colonise sediments but, this approach has some limitations. First, animals are isolated from biological interactions such as predation, permitting unrealistic behaviours. For example, the hunger levels of animals can impact upon their activity levels and, thus, sediment transport (Statzner *et al.*, 1996; Zanetell & Peckarsky, 1996). Second, artificial sediment beds rarely replicate the structural complexity of natural sediment beds. In addition, sediment grain-size, structure and organic content, as well as the hydrodynamic and thermal environment, all influence the burrowing activity of animals (discussed in section 7.4.2) and, consequently, their impact on sediment stability. Some studies have transferred natural sediments to the laboratory. For example, cores taken from the Skeffling mudflat (Humber estuary, UK) were used to measure the entrainment threshold of sediments under controlled laboratory conditions (Widdows *et al.*, 2000a). Although this ensures that the sediments are representative of the natural environment, there are still problems with this approach. In particular it is important to ensure that sediment is not disturbed during removal from the field, transportation and installation in the hydraulic facility. Disturbance during translocation will have implications for the entrainment stresses required to mobilise sediments. In fine grained systems it is also important to use water obtained from the field site because water chemistry, including salinity, exerts an important control on the erosion of cohesive sediments (Winterwerp & van Kesteren, 2004).

7.7 CONCLUSIONS

Benthic invertebrates (macrozoobenthos) are ubiquitous in most aquatic environments and can have large and pervasive impacts on the physical environment. Most previous research has concentrated on the impacts of burrowing taxa on sediment mixing and mobilisation, but it is apparent that organisms have a substantial topographic impact, particularly on fine-grained marine substrates; this will influence the near-bed hydraulic environment. The details of these phenomena and their relative significance remain largely unknown and understudied. Animals also alter near-bed hydrodynamics by generating currents to irrigate their burrows. It is known that the volume of water processed in this way can be large and is significant for biogeochemical cycling, but the discharge of these flows and their impacts on boundary-layer hydraulics remain unknown and may be of significance in areas with large colonies of organisms. In general, less is known about the interactions between flow hydraulics, sediment transport and biota in freshwater environments than in marine environments.

Experiments focused on the interaction between animals and their environment must account for, and in many cases replicate, the environmental and ecological context of the species and process being studied. For example, it is important that sediment alterations by animals are studied in conditions equivalent to field environments where the organism naturally occurs. Similarly, the hydraulic environment and thermal regime will control animal behaviour and therefore need to be considered in experimentation. Of equal importance are ecological interactions between animals that control the activity levels. For example, an animal is likely to be less active when a predator is present and more active in the presence of food. Therefore, animal densities in experiments should approximate natural densities in the field. Even in highly controlled, well-designed experiments, results will be highly variable due to the inherent variability in animal activity between individuals and of an individual through time. This needs to be acknowledged and, as a consequence, replication of experiments is important. The variable nature of interactions between environments and animals also makes extrapolation to larger scales difficult because results derived from a few animals over a short time period may not be representative of all individuals in that species, or activity throughout the year. For example, the seasonality of many animals in temperate latitudes results in their impact being limited to warmer months and all laboratory experiments would therefore have to be carried out either in summer or under temperature controlled conditions.

Chapter 8

Conclusion: Decision-making framework

R.E. Thomas, S.J. McLelland, M.F. Johnson,
L.E. Frostick, S.P. Rice, W.E. Penning & J.T. Dijkstra

8.1   INTRODUCTION

Section 2 of this book has identified the potential for physical models of ecohydraulics to bridge the gaps between observations of the natural, present-state, environment and theoretical, stochastic and numerical models. The integration of these research fields will improve our ability predict the impact of current and future changes in boundary conditions on aquatic ecology. Physical modelling in ecohydraulics may also be essential to investigate problems that cannot be studied in the field. This is often due to the inability to control boundary conditions in the field, but for such experiments to be meaningful, care has to be taken to ensure both the ecology and hydraulics are properly represented (See Chapters 5–7, this book).

Section 1 of this book outlined the decision-making process required for effective ecohydraulics experimentation. The corollary of this decision-making process is the need for a "nested approach" where results from field studies, laboratory experiments and numerical models are used in parallel to address the objectives of any study. Each of these approaches to studying ecohydraulics is essential to informing and refining the results of the other fields of study. For example, data from the natural environment are required to ensure that physical models reproduce behaviour found in the wild, irrespective of whether living or surrogate plants or animals are used; numerical models should be validated using field and laboratory data; and physical models enable direct observation of organism behaviour which may not be possible otherwise. Adopting this nested approach will enhance the quality of interpretations and, in turn, this will lead to more robust conclusions.

The need for a nested approach means that ecohydraulics has to be an interdisciplinary and multidisciplinary field involving engineers, applied mathematicians, geomorphologists, microbiologists, ecologists and fluid dynamicists. It is to be expected, therefore, that scientists in these disparate fields have different knowledge bases and ask fundamentally different questions (see Rice *et al.*, 2010a and references therein). For studies in ecohydraulics to be successful, researchers from different fields need to communicate in a manner that is unambiguous, specific and jargon-free. Nevertheless, it is clear that there are likely to be significant differences between the aims and objectives of studies formulated within each discipline. In utilising this book to design your own ecohydraulic experiment, the chapters from which you have gained most knowledge are likely to have been dependent upon your area of expertise. We anticipate that scientists from a biological or ecological background may not have previously considered the following key questions:

- Is the experimental facility large enough to conduct experiments at full scale? Does your model need to be scaled in order to adequately represent your prototype? Which scale modelling approach will you adopt in order to prevent process distortion?
- What is the grain-size distribution, density, structure and geochemistry of your prototype sediment? Can you source similar sediment for your experiment, or if you are conducting a scaled experiment, can you source sediment that scales appropriately?
- What are the temperature, turbidity and geochemistry of your prototype water? Can you source and maintain similar water properties for your experiment?
- What flow properties will be measured and is the available equipment appropriate? Will you need to add seeding material to ensure high quality measurements?
- What hydrodynamic conditions occur in your prototype? Are there waves? Are currents unidirectional or bi-directional? If the current varies spatio-temporally, how will you adequately characterise that variability in your experiment? How long will you sample for, at what rate and over what spatial area?

Many of these questions have been discussed within Chapters 2, 3, and 4 of this book. Conversely, if you are a geomorphologist, earth scientist or hydraulic engineer, you may not have previously considered the following key questions:

- Can the natural environment be adequately simulated within the experimental facility? Is it possible to reproduce the required temperature, lighting and biogeochemical conditions?
- Will the experiment be undertaken with live organisms, live surrogate organisms or surrogate organisms?
- If you are using live organisms of any type, can you source those organisms? What acclimatisation facilities will be needed? What will be the condition of the organism prior to experiments? Will they be hungry, tired, or stressed? What

steps need to be undertaken to ensure organism well-being during experiments? What abiotic factors need to be controlled in order to prevent unnatural organism behaviours and physiologies?

Regardless of the background and expertise of researchers, questions remain over the extent to which physical experiments can reproduce the natural field environment since, by their very nature and purpose, physical experiments involve some reduction in complexity. Therefore careful consideration must be given to the processes that will be represented within the physical experiment. This book has sought to provide a framework by which researchers can make the decisions needed to design an ecohydraulic experiment. In this chapter, we attempt to make this framework more explicit by providing a decision-making matrix. With careful consideration of these issues, researchers from the disparate disciplines involved will be better able to identify the key physico-chemical processes and organism characteristics from the field that can be represented in the physical experiments and to understand whether the steady or stable laboratory conditions can be extrapolated to unsteady or variable field conditions. The matrix addresses these issues by focussing on the following questions:

- What is the focus of the research?
- What type of experiment?
- What type of experimental facility will be used? (note that the available facility may not be most appropriate)
- What is the scale of the experiments?
- Will conditions, parameters and/or results need to be scaled?
- How long will experiments with living organisms last?
- What parameters will be measured?
- Will living animals or surrogates be used?
- What organism taxa will be used?

The decision-making framework offers an alternative means of navigating this book and methods of determining important considerations before embarking on experimentation at the interface of ecology and environmental science in aquatic environments. In the left-hand column a question is posed, the answer of relevance to your study can be found in the column to the right of the question. The next column lists considerations of significance to that response and in the column furthest to the right, is a reference to the chapter(s) of most relevance to that point where more detail, and additional references, can be found. Whilst some decisions will be forced upon scientists such as the equipment or facility that are available for use, the decision-making framework will help ensure that such compromises can be made whilst still being able to successfully explore the hypotheses to be investigated.

8.2 DECISION-MAKING FRAMEWORK

Considerations when planning experiments at the interface between biological and environmental science

Question	Answer	Important considerations	Chapter(s)
What is the focus of the research?	Drag, roughness and detailed hydrodynamics around organisms	Flexibility will influence drag forces and might influence measurements due to movement.	6
		Animal behaviour (posture, position, orientation) will influence any impact and will be partially determined by husbandry.	2, 7
		How to attach the organism to a drag force meter?	4, 7
		Will the results be used for up-scaling to larger applications? – What is required for successful scaling?	3
		Flow measurements over sufficient duration to identify turbulence.	4
		Transient conditions near leading edge of patch or semi-stationary (fully adapted) conditions over extended spatial scale?	4
	Biologically-generated flows	Requires sensitive measurement devices (LDA; PIV).	4
		Food presence, quality and quantity will influence feeding flows.	2
		Animal movement and daily cycles of behaviour might influence results and up-scaling.	2
	Sediment stability and mixing	Source, structure and composition of sediment.	7
		Presence of organic matter may induce disturbance and digging.	2, 7
		Length of time that organisms are left to occupy sediments.	7
		Sediment erosion is a one-time event. Can you replicate outcome?	Section 2, all chapters
	Wave dampening, flow alterations by patches of organisms and bioprotection	Attachment strength may be dependent on flow history (as for mussels).	2
		Where do you place your measurement equipment and what is the frequency of measurements?	4
		How do you parameterise your patch?	4
		Could you use surrogates instead of living organisms?	3
	Link to modelling	As above, plus:	
		What results do you need?	3
		Is scaling of results required and how will this be achieved?	
		Requires large volumes of appropriate sediment and water.	2
		Are surrogates or scaling required?	3

What type of experiment?	In situ flume in the field Flume	Side-wall drag.		N/A
		Does the flume tilt and what is the slope?		2
		Is the flume large enough for the experiments in terms of abiotic conditions to be generated and space for organism?		2
		Side-wall drag and wave reflection.		2
		Not all flumes can take saltwater as it corrodes metal.		2
		Flumes can only generate a limited range of flows which may limit applications.		2
		Are there any pollutants in the water? (if multiple flumes circulate through a single tank there is a need to take into account other experiments)		2
What type of experimental facility will be used? (note that the available facility may not be most appropriate)	Unidirectional	No multi-directional flow. May not be able to produce waves. Side-wall effects.	Frostick et al. (2011) chapter 1	2
	Annular and racetrack	Is it of sufficient length for creating stable flow conditions? Reduced boundary layer. Circulation of sediments.		
	Wave basin	Wave reflection, slope of beach. How many gauges and what is their placement in relation to each other?	Frostick et al. (2011) chapter 2	
What is the scale of the experiments?	Small scale i.e. <0.1 m	Is the measurement equipment sensitive enough?		4
		Error associated with unnatural animal behaviour highly significant (in comparison to larger experiments using many animals). Movement might affect measurement opportunities.		2
	Intermediate scale i.e. 0.1 m–10 m	Using multiple individuals – competition/density effects?		2
		Are multiple species required?		2
		Does a unidirectional flume sufficiently represent the natural system you are interested in (e.g. flow around patches might be affected by side-wall effects)?	Frostick et al. (2011) chapter 1	
	Large scale >10 m	Requires large stock of organisms (where collected, kept and reared)?		2
Will conditions, parameters and/or results need to be scaled?	No – the experiments are 1:1 scale	Is the flume of sufficient size for 1:1 scale biological processes, such as animal movement, plant growth?		2, 5, 6, 7

(Continued)

(Continued).

Question	Answer	Important considerations	Chapter(s)
	Yes	Is the flume of sufficient size and power to generate 1:1 scale abiotic processes, such as sediment grain-sizes and hydraulic forces?	Frostick et al. (2011) chapter 1
		How can abiotic conditions be scaled (i.e. sediment (grain Reynolds number) and hydraulic (Froude number, Reynolds number, Weber number) scaling)?	3
		How can biological parameters be scaled?	3
		Which biological parameters be scaled (elasticity, buoyancy)?	3
How long will experiments with living organisms last?	<2 hours	Water source and sediment source.	2
		Pollutants (heavy metals) from the pump and/or pipes.	2, 5
		Viruses, bacteria and chemicals in the water and sediment.	2, 5
		Water temperature and oxygen levels.	2, 5
		Noise and vibrations.	2
		Animal conditioning and state before experiments.	2
		Light levels (e.g. if studying nocturnal animals; shadows may disturb animals).	2, 5
	2 hour–12 hours	As above, plus:	
		Controlling light levels (day-night cycles).	2, 5
		Food or nutrient levels (quantity, quality and how provided).	2, 5
	12 hours–2 days	As above, plus:	
		Maintaining oxygen levels.	2, 5
		Controlling temperature (despite increasing temperature due to running pump(s)).	2, 5
		Timing of feeding.	
	>7 days	As above, plus:	2
		Seasonality of behaviour and/or physiology.	2, 7
		Growth of plants might affect their density and biomass distribution.	2, 5, 6
		Feeding cycles.	2, 7
		May need to include biological interactions to ensure and maintain morphology and behaviour that is similar to wild equivalents.	2, 5, 6, 7
		Need to account for changing density and biomass of organisms due to growth, birth and death rates?	2, 5, 6, 7

What parameters will be measured?	Drag and lift	Are the force transducers sensitive enough to measure drag on the organism?	4
		Is it turbulent fluctuations or mean drag that is of importance?	4
	Velocity, turbulence and bed shear stress	Compromise of scale (small-scale and detailed using, for instance, PIV or large-scale, averaged point measures using ADVs, EMS, etc.).	4
		Is seeding material required? (Depending on the choice of material, it may alter pH and conductivity)	4
		Complexity of post-processing (PIV complicated; ADV relatively easy).	4
		Multi-point or single point measurements (spatial scale)?	4
		Measurements will need to last long enough to adequately characterise turbulence.	4
		Use sensors in the bed or derive shear stress from flow velocity profile?	4
	Sediment stability	How will sediment stability be quantified (transport rate, entrainment threshold, etc)?	Frostick et al. (2011) chapter 4
		Sediment source, structure, composition, grain-size distribution.	
		What velocities will be used and can the flume generate them?	6
	Sediment mixing	What depth of sediment is required?	7
		Sediment source, structure, composition, grain-size distribution.	7
		How will mixing be quantified (X-ray, MRI, resin casts)?	5
	Biological parameters	What will be measured (i.e. size, biomass, density, flexibility)?	
		How will variability in biotic parameters between individuals and between replications be accounted for?	6,7
Will living animals or surrogates be used?	Living organisms	Is the facility large enough to house the organism?	2
		Will you need licences to use the organism (i.e. is it an invasive or rare species) and do you need training to handle the organism?	2
		Organism source (raised in lab, field collected) will affect behaviour and/or physiology.	2
		Will the organism be acclimatised to laboratory conditions?	2
		Animal behaviour is controlled by biological interactions that are not replicated in experiments with one individual or one species.	2
		What condition of organism will be used (i.e. hungry, stressed, injured, reproductive state, standardised size)?	2

(Continued)

(Continued).

Question	Answer	Important considerations	Chapter(s)
		Relative height of biota in comparison to the water depth will affect relative significance of hydrodynamic effects.	2, 6
		Relative width of biota in comparison to the channel width will affect relative significance of hydrodynamic effects.	
		Are you interested in the natural condition of organisms (i.e. with attached organisms, injuries, pocks, etc.)?	7
	Surrogate	What is the scale of study – how precise does the surrogate need to be?	3
		How will the surrogate be made?	3
		What materials will be used?	3
		What are the important conditions to be mimicked (i.e. flexibility, surface roughness; leaves per stem; structures per m³ or m²)?	3
		How will these conditions, and the conditions of the surrogate, be parameterised (e.g. Young's modulus, flexural rigidity, porosity, buoyancy)?	3
		How will the success of the surrogate be validated?	3
	Scaled, biological surrogate species	Does the surrogate mimic larger equivalents?	3
		How long does the surrogate take to grow?	3
		How do you scale biological properties such as Young's modulus, flexural rigidity, porosity, buoyancy?	3
		Necessary to also scale abiotic conditions i.e. sediment and flow conditions.	3, Frostick et al. (2011) chapter 4
What organism taxa will be used?	Plants	What sediment will be used? – need nutrients, stability and certain sizes.	2
		Water source, chemistry and salinity important for health and will affect buoyancy.	2
		Are there any pollutants in the water or sediment?	2
		Need light to grow and remain healthy.	2

	How will plant be attached to substrate (in sediment, in pots or tied to a grid or rack)?	2
	What size plant will be used relative to water depth and what properties are important?	2
	Age and condition of the plant will affect their physiology, strength and buoyancy.	2
	Buoyancy can be affected by photosynthesis (attached gas bubbles).	2
Biofilms	*As for plants*, plus:	
	The environment will impact biofilm morphology and texture.	5
	Both autotrophic and heterotrophic organisms need to be catered for in terms of food supply.	2
Sessile animals	What and when will the organism be fed?	2
	What water temperature is required?	2
	What oxygen levels need to be maintained?	2
	Need to account for pollutants and chemicals in water and sediment?	2
	Is there a need to replicate biological interactions (i.e. predation, competition) to ensure natural behaviours?	2
Mobile animals	How will animals be attached to the bed?	4, 7
	As for sessile animals, plus:	2
	Require space to move around and explore.	2
	Need to replicate interactions with other individuals.	2

References

Abdeen MAM. 2008. Predicting the impact of vegetations in open channels with different distributaries' operations on water surface profile using artificial neural networks. *Journal of Mechanical Science and Technology* **22**: 1830–1842.

Abdelrhman MA. 2007. Modeling coupling between eelgrass *Zostera marina* and water flow. *Marine Ecology Progress Series* **338**: 81–96.

Ackerman JD, Hoover TM. 2001. Measurement of local bed shear stress in streams using a Preston-static tube. *Limnology and Oceanography* **46**: 2080–2087.

Ackerman JD, Wong L, Ethier CR, Allen DG, Spelt JK. 1994. Preston-static tubes for the measurement of wall shear stress. *Journal of Fluids Engineering* **116**: 645–649.

Adrian RJ. 1996. Laser Velocimetry. In: Goldstein RJ. (ed) *Fluid Mechanics Measurements*. 2nd Edition Taylor & Francis Ltd., London, UK. pp. 175–299.

Alekseev VR, Starobogatov YI. 1996. Types of diapause in Crustacea: definitions, distribution, evolution. *Hydrobiologia* **320**: 15–26.

Allan IJ, House WA, Parker A, Carter JE. 2004. Transport and distribution of lindane and simazine in a riverine environment: measurements in bed sediments and modelling. *Pest Management Science* **60**: 417–433.

Allan IJ, House WA, Parker A, Carter JE. 2005. Diffusion of the synthetic pyrethroid permethrin into bed-sediments. *Environmental Science and Technology* **39**: 523–530.

Allanson BR, Skinner D, Imberger J. 1992. Flow in prawn burrows. *Estuarine Coast Shelf Science* **35**: 253–266.

Allen DC, Vaughn CC. 2009. Burrowing behavior of freshwater mussels in experimentally manipulated communities. *Journal of the North American Benthological Society* **28**: 93–100.

Allen DC, Vaughn CC. 2011. Density-dependent biodiversity effects on physical habitat modification by freshwater bivalves. *Ecology* **92**: 1013–1019.

Aller RC, Cochran JK. 1976. ^{234}Th/^{238}U disequilibrium in near-shore sediment: particle reworking and diagenetic time scales. *Earth and Planetary Science Letters* **29**: 37–50.

Aller RC, Yingst JY. 1980. Relationships between microbial distributions and the anaerobic decomposition of organic matter in surface sediments of Long Island Sound USA. *Marine Biology* **56**: 29–42.

Aller RC. 1988. Benthic fauna and biogeochemical processes in marine sediments: the role of burrow structures. In: Blackburn TH, Sørensen J. (eds) *Nitrogen cycling in coastal marine environments* John Wiley and Sons, Chichester, UK. pp. 301–338.

Amos CL, Sutherland TF, Cloutier D, Patterson S. 2000. Corrasion of a remoulded cohesive bed by saltating littorinid shells. *Continental Shelf Research* **20**: 1291–1315.

Amyot J-P, Downing JA. 1997. Seasonal variation in vertical and horizontal movement of the freshwater bivalve *Elliptio complanata* (Mollusca: Unionidae). *Freshwater Biology* **37**: 345–354.

Amyot J-P, Downing JA. 1998. Locomotion in *Elliptio complanata* (Mollusca: Unionidae): a reproductive function? *Freshwater Biology* **39**: 351–358

Andersen TJ. 2001a. The role of fecal pellets in suspended sediment settling velocities at an intertidal mudflat, the Danish Wadden sea. In: McAnally WH, Mehta AJ. (eds) *Coastal and Estuarine Fine Sediment Processes*. Elsevier Science, Amsterdam, The Netherlands. 387–401.

Andersen TJ. 2001b. Seasonal variation in erodability of two temperate, microtidal mudflats. *Estuarine, Coastal and Shelf Science* **53**: 1–12.

Andersen TJ, Jensen KT, Lund-Hansen L, Mouritsen KN, Pejrup M. 2002. Enhanced erodibility of fine-grained marine sediments by *Hydrobia ulvae*. *Journal of Sea Research* **48**: 51–58.

Antonarakis AS, Richards KS, Brasington J, Bithell M. 2009. Leafless roughness of complex tree morphology using terrestrial lidar. *Water Resources Research* **45**: W10401.

Aranguiz R, Villagran M, Eyzaguirre G. 2011. Use of trees as a tsunami natural barrier for Concepcion, Chile. *Journal of Coastal Research* **I64**: 450–454.

Armanini A, Righetti M, Grisenti P. 2005. Direct measurement of vegetation resistance in prototype scale. *Journal of Hydraulic Research* **43**: 481–487.

Arnon S, Gray KA, Packman AI. 2007. Biophysicochemical process coupling controls nitrate use by benthic biofilms. *Limnology and Oceanography* **52**: 1665–1671.

Arnon S, Marx LP, Searcy KE, Packman AI. 2010. Effects of overlying velocity, particle size, and biofilm growth on stream–subsurface exchange of particles. *Hydrological Processes* **24**: 108–114.

Asaeda T, Rajapakse L, Kanoh M. 2010. Fine sediment retention as affected by annual shoot collapse: *Sparganium erectum* as an ecosystem engineer in a lowland stream. *River Research and Applications* **26**: 1153–1169.

Asano T, Deguchi H, Kobayashi N. 1992. Interaction between water waves and vegetation, 23rd Coastal Engineering Conference, ASCE, Reston, VA, pp. 2710–2723.

Asselman NM, Middelkoop H, Ritzen MR, Straatsma MW. 2002. Assessment of the hydraulic roughness of river floodplains using laser altimetry. Selected papers for IAHS International Symposium: The structure, function and management implications of fluvial sedimentary systems, 2nd–6th September 2002, Alice Springs, Australia. pp. 381–388.

Astall CM, Taylor AC, Atkinson RJA. 1997. Behavioural and physiological implications of a burrow-dwelling lifestyle for two species of Upogebiid mud-shrimp (Crustacea: Thalassinidea). *Estuarine, Coastal and Shelf Science* **44**: 155–168.

Atkinson RJA, Chapman CJ. 1984. Resin casting: a technique for investigating burrows sublittoral sediments. *Progress in Underwater Science* **10**: 109–115.

Atkinson RJA, Taylor AC. 1991. Burrows and burrowing behaviour of fish. *Symposium of the Zoological Society of London* **63**: 133–155.

Augspurger C, Kusel K. 2010. Flow velocity and primary production influences carbon utilization in nascent epilithic stream biofilms. *Aquatic Science* **72**: 237–243.

Augustin LN, Irish JL, Lynett P. 2009. Laboratory and numerical studies of wave damping by emergent and near-emergent wetland vegetation. *Coastal Engineering* **56**: 332–340.

Bailey-Brock JH. 1984. Ecology of the tube-building polychaete *Diopatra leuckarti* Kinberg, 1865 (Onuphidae) in Hawaii: community structure, and sediment stabilizing properties. *Zoological Journal of the Linnean Society* **80**: 191–199.

Baker B, Olszyk DM, Tingey D. 1996. Digital image analysis to estimate leaf area. *Journal of Plant Physiology* **148**: 530–535.

Banta GT, Andersen O. 2003. Bioturbation and the fate of sediment pollutants. *Vie Millieu* **53**: 233–248.

Baptist MJ, Babovic V, Rodríguez Uthurburu J, Keijzer M, Uittenbogaard RE, Mynett A, Verwey A. 2007. On inducing equations for vegetation resistance. *Journal of Hydraulic Research* **45**: 435–450.

Baptist MJ, Haasnoot M, Cornelissen P, Icke J, van der Wedden G, de Vriend HJ, Gugic G. 2006. Flood detention, nature development and water quality along the lowland river Sava, Croatia. *Hydrobiologia* **565**: 243–257.

Baptist MJ, Penning WE, Duel H, Smits AJM, Geerling GJ, Van der Lee G, Van Alphen JSL. 2004. Assessment of cyclic floodplain rejuvenation on flood levels and biodiversity in the Rhine river. *River Research and Applications* **20**: 285–297.

Baptist MJ, van den Bosch LV, Dijkstra JT, Kapinga S. 2005. Modelling the effects of vegetation on flow and morphology in rivers. *Archiv Fur Hydrobiologie* **155**: 339–357.

Barkdoll BD. 2002. Discussion of "Mean Flow and Turbulence Structure of Open-Channel Flow Through Non-Emergent Vegetation" by Fabián López and Marcelo H. García. *Journal of Hydraulic Engineering, ASCE* **128**: 1032.

Barko JW, Adams MS, Clesceri NL. 1986. Environmental factors and their consideration in the management of submerged aquatic vegetation: a review. *Journal of Aquatic Plant Management* **24**: 1–10.

Barko JW, Gunnison D, Carpenter SR. 1991. Sediment interacters with submerged macrophyte growth and community dynamics. *Aquatic Botany* **41**: 41–65.

Barko JW, Hardin DG, Matthews MS. 1982. Growth and morphology of submersed freshwater macrophytes in relation to light and temperature. *Canadian Journal of Botany* **60**: 877–887.

Barko JW, Smart RM. 1986. Sediment-related mechanisms of growth limitation in submersed macrophytes. *Ecology* **67**: 1328–1340.

Basil J, Sandeman D. 2000. Crayfish (*Cherax destructor*) use tactile cues to detect and learn topographical changes in their environment. *Ethology* **106**: 247–259.

Battin TJ, Butturini A, Sabater F. 1999. Immobilization and metabolism of dissolved organic carbon by natural sediment biofilms in a Mediterranean and temperate stream. *Aquatic Microbial Ecology* **19**: 297–305.

Battin TJ, Kaplan LA, Newbold JD, Cheng X, Hansen C. 2003. Effects of current velocity on the nascent architecture of stream microbial biofilms. *Applied and Environmental Microbiology* **69**: 5443–5452.

Battin TJ, Sengschmitt D. 1999. Linking sediment biofilms, hydrodynamics, and river bed clogging: evidence from a large river. *Microbial Ecology* **37**: 185–196.

Beaumont AR, Wei JHC. 1991. Morphological and genetic variation in the Antarctic limpet *Nacella concinna* (Strebel, 1908). *Journal of Molluscan Studies* **57**: 443–450.

Belanger SE, Meiers EM, Bausch RG. 1995. Direct and indirect ecotoxicological effects of alkyl sulfate and alkyl ethoxysulfate on macroinvertebrates in stream mesocosms. *Aquatic Toxicology* **33**: 65–87.

Bell EC, Gosline JM. 1997. Strategies for life in flow: tenacity, morphometry, and probability of dislodgment of two *Mytilus* species. *Marine Biology Progress Series* **159**: 197–208.

Bell SS. 1985. Habitat complexity of polychaete tube-caps: influence of architecture on dynamics of a meioepibenthic assemblage. *Journal of Marine Research* **43**: 647–671.

Bender K, Davis WR. 1984. The effect of feeding by *Yoldia limatula* on bioturbation. *Opheila* **23**: 91–100.

Bennett SJ, Bridge JS, Best JL. 1998. The fluid and sediment dynamics of upper-stage plane beds. Journal of Geophysical Research **103**: 1239–1274.

Bennett SJ, Pirim T, Barkdoll BD. 2002. Using simulated emergent vegetation to alter stream flow direction within a straight experimental channel. *Geomorphology* **44**: 115–126.

Bennett SJ, Wu W, Alonso CV, Wang S. 2008. Modeling fluvial response to in-stream woody vegetation: Implications for stream corridor restoration. *Earth Surf. Process. Landforms* **33**: 890–909.

Benninger LK, Aller RC, Cochran JK, Turekian KK. 1979. Effects of biological sediment mixing on the ^{210}Pb chronology and trace metal distribution in a Long Island Sound sediment core. *Earth and Planetary Science Letters* **43**: 241–259.

Bertoldi W, Gurnell AM, Surian N, Tockner K, Zanoni L, Ziliani L, Zolezzi G. 2009. Understanding reference processes: linkages between river flows, sediment dynamics and vegetated landforms along the Tagliamento River, Italy. *River Research and Applications* 25: 501–516.

Besemer K, Hödl I, Singer G, Battin TJ. 2009b. Architectural differentiation reflects bacterial community structure in stream biofilms. *The ISME Journal* 3: 1318–1324.

Besemer K, Singer G, Hödl I, Battin TJ. 2009a. Bacterial community composition of stream biofilms in spatially variable-flow environments. *Applied and Environmental Microbiology* 75: 7189–7195.

Besemer K, Singer G, Limberger R, Chlup A-K, Hochedlinger G, Hödl I, Baranyi C, Battin TJ. 2007. Biophysical controls on community succession in stream biofilms. *Applied and Environmental Microbiology* 73: 4966–4974.

Best EPH, Boyd WA. 2003. *A simulation model for growth of the submerged aquatic macrophyte sago pondweed (*Potamogeton pectinatus *L.).* US Army Corps of Engineers, Washington DC, USA.

Beukema JJ. 1982. Calcimass and carbonate production by molluscs on the tidal flats in the Dutch Wadden Sea: II The edible cockle, *Cerastoderma edule. Netherlands Journal of Sea Research* 15: 391–405.

Biggs BJF. 1996. Patterns in benthic algae of streams. In: Stevenson RJ, Bothwell ML, Lowe RL. (ed) *Algal Ecology: Freshwater Benthic Ecosystems.* Academic, San Diego, USA. pp. 31–56.

Biggs BJF, Duncan MJ, Francoeur SN, Meyer WD. 1997. Physical characterisation of microform bed cluster refugia in 12 headwater streams, New Zealand. *New Zealand Journal of Marine and Freshwater Research* 31: 413–422.

Biggs BJF, Goring DG, Nikora VI. 1998. Subsidy and stress responses of stream periphyton to gradients in water velocity as a function of community growth form. *Journal of Phycology* 34: 598–607.

Biggs BJF, Hickey CW. 1994. Periphyton responses to a hydraulic gradient in a regulated river in New Zealand. *Freshwater Biology* 32: 49–59.

Biggs BJF, Thomsen HA. 1995. Disturbance of stream periphyton by perturbations in shear stress: time to structural failure and differences in community resistance. *Journal of Phycology* 31: 233–241.

Biles CL, Paterson DM, Ford RB, Solan M, Raffaelli DG. 2002. Bioturbation, ecosystem functioning and community structure. *Hydrology and Earth System Sciences* 6: 999–1005.

Bintz JC, Nixon SW. 2001. Responses of eelgrass *Zostera marina* seedlings to reduced light. *Marine Ecology Progress Series* 223: 133–141

Birkett C, Tollner EW, Gattie DK. 2007. Total suspended solids and flow regime effects on periphyton development in a laboratory channel. *Transactions of the American Society of Agricultural and Biological Engineers* 50: 1095–1104.

Blanchard GF, Guarini JM, Richard P, Gros P, Mornet F. 1996. Quantifying the short-term temperature effect on lightsaturated photosynthesis on intertidal microphytobenthos. *Mar Ecol. Prog. Ser.,* 134: 309–313.

Blanchard GF, Sauriau P-G, Cariou-Le Gall V, Gouleau D, Garet M-J, Olivier F. 1997. Kinetics of tidal resuspension of microbiota: testing the effects of sediment cohesiveness and bioturbation using flume experiments. *Marine Ecology Progress Series* 151: 17–25.

Blanckaert K. 2009. Saturation of curvature-induced secondary flow, energy losses, and turbulence in sharp open-channel bends: Laboratory experiments, analysis, and modelling. *Journal of Geophysical Research, Earth Surface* 114: F03015. DOI:10.1029/2008 JF001137.

Blanckaert K. 2010. Topographic steering, flow recirculation, velocity redistribution, and bed topography in sharp meander bends. *Water Resources Research* 46: W09506. DOI:10.1029/2009 WR008303.

Blanckaert K, de Vriend HJ. 2004. Secondary flow in sharp open-channel bends. *Journal of Fluid Mechanics* **498**: 353–380.

Blanckaert K, de Vriend HJ. 2005a. Turbulence characteristics in sharp open-channel bend. *Physics of Fluids* **17**(5): 055102, DOI:10.1063/1.1886726.

Blanckaert K, de Vriend HJ. 2005b. Turbulence structure in sharp open-channel bends. *Journal of Fluid Mechanics* **536**: 27–48.

Blanckaert K, Graf WH. 2001. Experiments on flow in an open-channel bend: Mean flow and turbulence. *Journal of Hydraulic Engineering, ASCE* **127**: 835–847.

Blanckaert K, Graf WH. 2004. Momentum transport in sharp open-channel bends. *Journal of Hydraulic Engineering, ASCE* **130**: 186–198.

Blanckaert K, Lemmin U. 2006. Means of noise reduction in acoustic turbulence measurements. *Journal of Hydraulic Research* **44**: 3–17.

Boeger MRT, Poulson ME. 2003. Morphological adaptations and photosynthetic rates of amphibious *Veronica anagallis-aquatica* L. (Scrophulariaceae) under different flow regimes. *Aquatic Botany* **75**: 123–135

Boller ML, Carrington E. 2007. Interspecific comparison of hydrodynamic performance and structural properties among intertidal macroalgae. *Journal of Experimental Biology* **210**: 1874–1884.

Bortolus A, Iribarne O. 1999. Effect of the SW Atlantic burrowing crab *Chasmagnathus granulata* on a *Spartina* saltmarsh. *Marine Ecology Progress Series* **178**: 79–88.

Bos AR, Bouma TJ, de Kort GLJ, van Katwijk MM. 2007. Ecosystem engineering by annual intertidal seagrass beds: Sediment accretion and modification. *Estuarine, Coastal and Shelf Science* **74**: 344–348.

Bottacin-Busolin A, Singer G, Zaramella M, Battin TJ, Marion A. 2009. Effects of streambed morphology and biofilm growth on the transient storage of solutes. *Environmental Science and Technology* **43**: 7337–7342.

Bottjer DJ, Hagadorn JW, Dornbos SQ. 2000. The Cambrian substrate revolution. *GSA Today* **10**: 1–7.

Botto F, Iribarne O. 1999. Effect of the burrowing crab *Chasmagnathus granulata* (Dana) on the benthic community of a SW Atlantic coastal lagoon. *Journal of Experimental Marine Biology and Ecology* **241**: 263–284.

Botto F, Iribarne O. 2000. Contrasting effects of two burrowing crabs (*Chasmagnathus granulata* and *Uca uruguayensis*) on sediment composition and transport in estuarine environments. *Estuarine, Coastal and Shelf Science* **51**: 141–151.

Bouchet VMP, Sauriau P-G, Debenay J-P, Mermillod-Blondin F, Schmidt S, Amiard J-C, Dupas B. 2009. Influence of the mode of macrofauna-mediated bioturbation on the vertical distribution of living benthic foraminifera: first insight from axial tomodensitometry. *Journal of Experimental Marine Biology and Ecology* **371**: 20–33.

Boulêtreau S, Garabétian F, Sauvage S, Sánchez-Pérez J-M. 2006. Assessing the importance of a self-generated detachment process in river biofilm models. *Freshwater Biology* **51**: 901–912.

Bouma TJ, de Vries MB, Herman PMJ. 2010. Comparing ecosystem engineering efficiency of 2 plant species with contrasting growth strategies. *Ecology* **91**: 2696–2704.

Bouma TJ, de Vries MB, Low E, Peralta G, Tánczos IC, van de Koppel J, Herman PMJ. 2005. Trade-offs related to ecosystem engineering: A case study on stiffness of emerging macrophytes. *Ecology* **86**(8): 2187–2199.

Bouma TJ, Friedrichs M, Klaasen P, van Wesenbeeck BK, Brun FG, Temmerman S, van Katwijk MM, Graf G, Herman PMJ. 2009b. Effects of shoot stiffness, shoot size and current velocity on scouring sediment from around seedlings and propagules. *Marine Ecology Progress Series* **388**: 293–297.

Bouma TJ, Friedrichs M, van Wesenbeeck BK, Temmerman S, Graf G, Herman PMJ. 2009a. Density-dependent linkage of scale-dependent feedbacks: a flume study on the intertidal macrophyte Spartina anglica. *Oikos* **118**: 260–268.

Bouma TJ, Olenin S, Reise K, Ysebaert T. 2009c. Ecosystem engineering and biodiversity in coastal sediments: posing hypotheses. *Helgoland Marine Research* **63**: 95–106.

Bouma TJ, van Duren LA, Temmerman S, Claverie T, Blanco-Garcia A, Ysebaert T, Herman PMJ. 2007. Spatial flow and sedimentation patterns within patches of epibenthic structures: Combining field, flume and modelling experiments. *Continental Shelf Research* **27**: 1020–1045.

Boylen CW, Sheldon RB. 1976. Submergent macrophytes: growth under winter ice cover. *Science* **194**: 841–842.

Bradding MG, Jass J, Lappin-Scott HM. 1995. Dynamics of bacterial biofilm formation. In: *Microbial biofilms*. Edited by Hilary Lappin-Scott, and J. William Costerton. Cambridge University Press, Cambridge, UK, pp. 46–63.

Bradley K, Houser C. 2009. Relative velocity of seagrass blades: Implications for wave attenuation in low-energy environments. *Journal of Geophysical Research* **114**: 1–13.

Bradshaw C. 1996. Bioturbation of reefal sediments by crustaceans in Phuket, Thailand. 8th International Coral Reef Symposium, Panama.

Braissant O, Decho AW, Dupraz C, Glunk C, Przekop KM, Visscher PT. 2007. Exopolymeric substances of sulfate-reducing bacteria: interactions with calcium at alkaline pH and implication for formation of carbonate minerals. *Geobiology* **5**: 401–411.

Branch GM, Marsh AC. 1978. Tenacity and shell shape in six *Patella* species: adaptive features. *Journal of Experimental Marine Biology and Ecology* **34**: 111–130.

Branch GM, Pringle A. 1987. The impact of the sand prawn Callianassa kraussi Stebbing on sediment turnover and on bacteria, meiofauna, and benthic microflora. *Journal of Marine Biology and Ecology* **107**: 219–235.

Braskerud BC. 2001. The influence of vegetation on sedimentation and resuspension of soil particles in small constructed wetlands. *Journal of Environmental Quality* **30**: 1447–1457.

Braudrick CA, Dietrich WE, Leverich GT, Sklar LS. 2009. Experimental evidence for the conditions necessary to sustain meandering in coarse-bedded river. *Proceedings of the National Academy of Science* **106**: 16936–16941.

Brayshaw AC, Frostick LE, Reid I. 1983. The hydrodynamics of particle clusters and sediment entrainment in coarse alluvial channels. *Sedimentology* **30**: 137–140.

Brennan A, McLachlan AJ, Wotton RS. 1979. Tubes and tube-building in a lotic chironomid (Diptera) community. *Hydrobiologia* **67**: 173–178.

Briggs KB, Williams KL, Richardson MD, Jackson DR. 2001. Effects of changing roughness on acoustic scattering: (1) Natural changes. In: Leighton TG. (ed) *Proceedings of the Institute of Acoustics-Acoustical Oceanography*. Bath University Press. pp. 375–382.

Bromley RG, Heinberg C. 2006. Attachment strategies of organisms on hard substrates: a palaeontological view. *Palaeogeography, Palaeoclimatology, Palaeoecology* **232**: 429–453.

Brönmark C, Hansson L-A. 2000. Chemical communication in aquatic systems: an introduction. *Oikos* **88**: 103–109.

Brown SL. 1998. Sedimentation on a Humber Saltmarsh. In: Black KS, Paterson DM, Cramp A. (eds) *Sedimentary Processes in the Intertidal Zone*. Special Publication 139. Geological Society, London. pp. 69–83.

Buchanan JB. 1963. The bottom fauna communities and their sediment relationships off the coast of Northumberland. *Oikos* **14**: 154–175.

Buckee C, Kneller BC, Peakall J. 2001. Turbulence structure in steady solute-driven gravity currents. In: McCaffrey WD, Kneller BC, Peakall J. (eds) *Particulate Gravity Currents*. Blackwell Science, Oxford, UK. pp. 173–188.

Buffin-Bélanger T, Roy AG. 1998. Effects of a pebble cluster on the turbulent structure of a depth-limited flow in a gravel-bed river. *Geomorphology* 25: 249–267.

Buffin-Bélanger T, Roy AG. 2005. 1 min in the life of a river: selecting the optimal record length for the measurement of turbulence in fluvial boundary layers. *Geomorphology* 68: 77–94.

Bulthuis DA, Brand GW, Mobley MC. 1984. Suspended sediments and nutrients in water ebbing from seagrass-covered and denuded tidal mudflats in a southern Australian embayment. *Aquatic Botany* 20: 257–266.

Burk RL, Lodge DM. 2002. Cued in: advances and opportunities in freshwater chemical ecology. *Journal of Chemical Ecology* 28: 1901–1917.

Butler DR. 1995. *Zoogeomorphology: animals as geomorphic agents.* Cambridge University Press, New York, USA.

Butman CA, Fréchette M, Geyer WR, Starczak VR. 1994. Flume experiments on food supply to the blue mussel *Mytilus edulis* L. as a function of boundary-layer flow. *Limnology and Oceanography* 39: 1755–1768.

Butterwick C, Heaney SI, Talling JF. 2005. Diversity in the influence of temperature on the growth rates of freshwater algae, and its ecological relevance. *Freshwater Biology* 50(2), 291–300.

Buttsworth DR, Elston SJ, Jones TV. 2000. Skin friction measurements on reflective surfaces using nematic liquid crystal. *Experiments in Fluids* 28: 64–73.

Cadée GC. 1976. Sediment reworking by *Arenicola marina* on tidal flats in the Dutch Wadden Sea. *Netherlands Journal for Sea Research* 10: 440–460.

Callaghan DP, Bouma TJ, Klaasen P, van der Wal D, Stive MJF, Herman PMJ. 2010. Hydrodynamic forcing on salt-marsh development: Distinguishing the relative importance of waves and tidal flows. *Estuarine, Coastal and Shelf Science* 89: 73–88.

Callaghan FM, Cooper GG, Nikora VI, Lamoroux N, Statzner B, Sagnes P, Radford J, Malet E, Biggs BJF. 2007. A submersible device for measuring drag forces on aquatic plants and other organisms. *New Zealand Journal of Marine and Freshwater Research* 41: 119–127.

Camfield FE. 1977. *Wind-wave propagation over flooded, vegetated land.* Coastal Engineering Research Center, U.S. Army Engineer Waterways Experiment Station, Vicksburg, Mississipi.

Canfield DE, Farquhar J. 2009. Animal evolution, bioturbation, and the sulfate concentration of the oceans. *Proceedings of the National Academy of Sciences* 106: 8123–8127.

Cardinale BJ, Gelmann ER, Palmer MA. 2004. Net spinning caddisflies as stream ecosystem engineers: the influence of Hydropsyche on benthic substrate stability. *Functional Ecology* 18: 381–387.

Carey DA. 1983. Particle resuspension in the benthic boundary layer induced by flow around polychaete tubes. *Canadian Journal of Fisheries and Aquatic Science* 40: 301–308.

Carrington E. 2002. Seasonal variation in the attachment strength of blue mussels: causes and consequences. *Limnology and Oceangraphy* 47: 1723–1733.

Cazelles B, Fontvieille D, Chau NP. 1991. Self-purification in a lotic ecosystem: a model of dissolved organic carbon and benthic microorganisms dynamics. *Ecological Modelling* 58: 91–117.

Cea L, Puertas J, Pena L. 2007. Velocity measurements on highly turbulent free surface flow using ADV. *Experiments in Fluids* 42: 333–348.

Cellino M, Graf WH. 1999. Sediment-laden flow in open-channels under noncapacity and capacity conditions. *Journal of Hydraulic Engineering, ASCE* 125: 455–462.

Chambers PA. 1987. Light and nutrients in the control of aquatic plant community structure. II. *In situ* observations. *Journal of Ecology* 75: 621–628.

Chang JS, Law R, Chang CC. 1997. Biosorption of lead, copper and cadmium by biomass of *Pseudomonas aeruginosa* PU21. *Water Research* 31: 1651–1658.

Chang HT, Rittmann BE, Amar D, Heim R, Ehrlinger O, Lesty Y. 1991. Biofilm detachment mechanisms in a liquid fluidized bed. *Biotechnology Bioengineering* **38**: 499–506.

Chanson H, Trevethan M, Koch C. 2007. Discussion of "Turbulence Measurements with Acoustic Doppler Velocimeters" by Carlos M. García, Mariano I. Cantero, Yarko Niño, and Marcelo H. García. *Journal of Hydraulic Engineering, ASCE* **133**: 1283–1286.

Chen S-C, Kuo Y-M, Li Y-H. 2011. Flow characteristics within different configurations of submerged flexible vegetation. *Journal of Hydrology* **398**: 124–134.

Chen S-N, Sanford LP, Koch EW, Shi F, North EW. 2007. A nearshore model to investigate the effects of seagrass bed geometry on wave attenuation and suspended sediment transport. *Estuaries and Coasts* **30**: 296–310.

Cheng NS. 2007. Power-law index for velocity profiles in open channel flows. *Advances in Water Resources* **30**: 1775–1784.

Choi SB, Yun YS. 2004. Lead biosorption by waste biomass of *Corynebacterium glutamicum* generated from lysine fermentation process. *Biotechnol Lett.* **26**: 331–336.

Choi SU, Kang HS. 2004. Reynolds stress modeling of vegetated open-channel flows. *Journal of Hydraulic Research* **42**: 3–11.

Choi SU, Yang W, Shin J. 2009. Laboratory experiments for 3D characteristics of depth-limited open-channel flows with submerged vegetation. In: *Water Engineering for a Sustainable Environment, Proceedings of the 33rd IAHR Congress, Vancouver BC, Canada, August 9–14, 2009*, IAHR: Madrid, Spain. pp. 462–471.

Chow VT. 1959. *Open Channel Hydraulics*. McGraw-Hill, New York, USA.

Christiansen T, Wiberg PL, Milligan TG. 2000. Flow and sediment transport on a tidal salt marsh surface. *Estuarine, Coastal and Shelf Science* **50**: 315–331.

Christy EJ, Sharitz RR. 1980. Characteristics of three populations of a swamp annual under different temperature regimes. *Ecology* **6**: 454–460.

Churchill SW. 1988. The Practical Use of Theory in Fluid Flows: Viscous Flows. Butterworth-Heinemann, Oxford.

Ciavola P. 2005. Sediment resuspension in the lagoon of Venice: short-term observations of natural and anthropogenic processes. *Zeitschrift für Geomorphologie Supplement Series* **141**: 1–15.

Ciraolo G, Ferreri GB, La Loggia G. 2006. Flow resistance of *Posidonia oceanica* in shallow water. *Journal of Hydraulic Research* **44**: 189–202.

Claret C, Fontvieille D. 1997. Characteristics of biofilm assemblages in two contrasted hydrodynamic and trophic contexts. *Microbial Ecology* **34**: 49–57.

Clement FE. 1916. Plant succession. Carnegie Institute Publication 242. Washington DC, USA.

Cloete TE, Westaard, D van Vuuren SJ. 2003. Dynamic response of biofilm to pipe surface and fluid velocity. *Water Science Technology* **47**(5), 57–59.

Coco G, Thrush SF, Green MO, Hewitt JE. 2006. Feedbacks between bivalve density, flow, and suspended sediment concentration on patch stable states. *Ecology* **87**: 2862–2870.

Cole BE, Thompson JK, Cloern JE. 1992. Measurement of filtration rates by infaunal bivalves in a recirculating flume. *Marine Biology* **113**: 219–225.

Coleman FC, Williams SL. 2002. Overexploiting marine ecosystem engineers: potential consequences for biodiversity. *Trends in Ecology and Evolution* **17**: 40–44.

Colin PL, Suchanek TH, McMurty G. 1986. Water pumping and particulate resuspension by Callianassids (Crustacea: Thalassinidea) at Enewetak and Bikini Atolls, Marshall Islands. *Bulletin of Marine Science* **38**: 19–24.

Collier KJ. 2002. Effects of flow regulation and sediment flushing on instream habitat and benthic invertebrates in a New Zealand river influenced by a volcanic eruption. *River Research and Applications* **18**: 213–226.

Commito JA, Celano EA, Celico HJ, Como S, Johnson CP. 2005. Mussels matter: postlarval dispersal dynamics altered by a spatially complex ecosystem engineer. *Journal of Experimental Marine Biology and Ecology* 316: 133–147.

Commito JA, Rusignuolo BR. 2000. Structural complexity in mussel beds: the fractal geometry of surface topography. *Journal of Experimental Marine Biology and Ecology* 255: 133–152.

Cooper GG, Callaghan FM, Nikora VI, Lamouroux N, Statzner B, Sagnes P. 2007. Effects of flume characteristics on the assessment of drag on flexible macrophytes and a rigid cylinder. *New Zealand Journal of Marine and Freshwater Research* 41: 129–135.

Cooper NJ. 2005. Wave dissipation across intertidal surfaces in the Wash tidal inlet, Eastern England. *Journal of Coastal Research* 21: 28–40.

Coops H, Geilen N, Verheij HJ, Boeters R, van der Velde G. 1996. Interactions between waves, bank erosion and emergent vegetation: an experimental study in a wave tank. *Aquatic Botany* 53: 187–198.

Corenblit D, Steiger J, Gurnell AM, Tabacchi E, Roques L. 2009. Control of sediment dynamics by vegetation as a key function driving biogeomorphic succession within fluvial corridors. *Earth Surface Processes and Landforms* 34(13), 1790–1810.

Cornelisen CD, Thomas FIM. 2006. Water flow enhances ammonium and nitrate uptake in a seagrass community. *Marine Ecology Progress Series* 312: 1–13.

Costerton JW. 1995. Overview of microbial biofilms. *Journal of Industrial Microbiology* 15: 137–140.

Cotton J, Wharton G, Bass J, Hepell C, Wotton R. 2006. The effects of seasonal changes to instream vegetation cover on patterns of flow and accumulation of sediment. *Geomorphology* 77: 320–334.

Crawford TW, Commito JA, Borowik AM. 2006. Fractal characterization of *Mytilus edulis* L. spatial structure in intertidal landscapes using GIS methods. *Landscape Ecology* 21: 1033–1044.

Creese R, Hooker S, De Luca S, Wharton Y. 1997. Ecology and environmental impact of *Musculista senhousia* (Mollusca: Bivalvia: Mytilidae) in Tamaki Estuary, Auckland, New Zealand. *New Zealand Journal of Marine and Freshwater Research* 31: 225–236.

Crimaldi JP, Koseff JR, Monismith SG. 2007. Structure of mass and momentum fields over a model aggregation of benthic filter feeders. *Biogeosciences* 4: 269–282.

Crimaldi JP, Thompson JK, Rosman H, Lowe, RJ, Koseff JR. 2002. Hydrodynamics of larval settlement: the influence of turbulent stress events at potential recruitment sites. *Limnology and Oceanography* 47: 1137–1151.

Crimes PT, Drosser ML. 1992. Trace fossils and bioturbation: the other fossil record. *Annual Review of Ecology and Systematics* 23: 339–360.

Crooks JA, Khim HS. 1999. Architectural vs. biological effects of a habitat-altering exotic mussel, *Musculista senhousia*. *Journal of Experimental Marine Biology and Ecology* 240: 53–75.

Crooks JA. 1996. The population ecology of an exotic mussel, *Musculista senhousia*, in a Southern California Bay. *Estuaries* 19: 42–50.

Crooks JA. 1998. Habitat alteration and community-level effects of an exotic mussel, *Musculista senhousia*. *Marine Ecology Progress Series* 162: 137–152.

Crosato A. 2008. *Analysis and modelling of River meandering*. PhD Thesis. Technical University of Delft. The Netherlands

Crossley MN, Dennison WC, Williams RR, Wearing AH. 2002. The interaction of water flow and nutrients on aquatic plant growth. *Hydrobiologia* 489: 63–70.

Cui J, Neary VS. 2008. LES study of turbulent flows with submerged vegetation. *Journal of Hydraulic Research* 46: 307–316.

Cuitat A, Widdows J, Pope ND. 2007. Effect of *Cerastoderma edule* density on near-bed hydrodynamics and stability of cohesive muddy sediments. *Journal of Experimental Marine Biology and Ecology* **346**: 114–126.

Cummings VJ, Thrush SF, Hewitt JE, Turner SJ. 1998. The influence of the pinnid bivalve *Atrina zelandica* (Gray) on benthic macroinvertebrate communities in soft-sediment habitats. *Journal of Experimental Marine Biology and Ecology* **228**: 227–240.

Cundy AB. Hopkinson L, Lafite R, Spencer K, Taylor JA, Ouddane B, Heppell CM, Carey PJ, Charman R, Shell D. 2005. Heavy metal distribution and accumulation in two Spartina sp.-dominated macrotidal salt marshes from the Seine estuary (France) and the Medway estuary (UK). *Applied Geochemistry* **20**: 1195–1208.

Cuny P, Miralles G, Cornet-Barthaux V, Acquaviva M, Stora G, Grossi V, Gilbert F. 2007. Influence of bioturbation by the polychaete *Nereis diversicolor* on the structure of bacterial communities in oil contaminated coastal sediments. *Marine Pollution Bulletin* **54**: 452–459.

D'Andrea AF, DeWitt TH. 2009. Geochemical ecosystem engineering by the mud shrimp *Upogebia pugettensis* (Crustacea: Thalassinidae) in Yaquina Bay, Oregon: density-dependent effects on organic matter remineralization and nutrient cycling. *Limnology and Oceanography* **54**: 1911–1932.

Daborn GR, Amos CL, Brylinsky M, Christian H, Drapeau G, Faas RW, Grant J, Long B, Paterson DM, Perillo GME, Piccolo MC. 1993. An ecological cascade effect: migratory birds affect stability of intertidal sediments. *Limnology and Oceanography* **38**: 225–231.

Dade WB, Hogg AJ, Boudreau BP. 2001. Physics of flow above the sediment-water interface, In: Boudreau BP, Jørgensen BB. (Eds) *The Benthic Boundary Layer: Transport Processes and Biogeochemistry*. Oxford University Press, Oxford, UK. pp. 4–43.

Darwin C. 1881. *The Formation of Vegetable Mould, through the Action of Worms, with Observations on their Habits*. John Murray, London, UK.

Dasilva EF, Almeida SFP, Nunes ML, Luis AT, Borg F, Hedlund M, De sà CM, Patinha C, Teixeira P. 2009. Heavy metal pollution downstream the abandoned Coval da Mó mine (Portugal) and associated effects on epilithic diatom communities. *Science of the Total Environment* **407**: 5620–5636.

Davey JT. 1994. The architecture of the burrow of *Nereis diversicolor* and its quantification in relation to sediment-water exchange. *Journal of Experimental Marine Biology and Ecology* **179**: 115–129.

Davidson MJ, Mylne KR, Jones CD, Phillips JC, Perkins RJ, Fung JCH, Hunt JCR. 1995. Plume dispersion through large groups of obstacles–a field investigation. *Atmospheric Environment* **29**: 3245–3256.

Davidson PA. 2004. *Turbulence: An Introduction for Scientists and Engineers*. Oxford University Press, Oxford, UK.

Davies-Colley RJ, Hickey CW, Quinn JM, Ryan PA.1992. Effects of clay discharges on streams 1. Optical properties and epilithon. *Hydrobiologia* **248**: 215–234.

Davis WR. 1993. The role of bioturbation in sediment resuspension and its interaction with physical shearing. *Journal of Experimental Marine Biology and Ecology* **171**: 187–200.

Day RT, Keddy PA, Mc Neill J, Carleton T. 1988. Fertility and disturbance gradients: a summary model for riverine marsh vegetation. *Ecology* **69**: 1044–1054

De Backer A, van Volen C, Vincx M, Degraer S. 2010. The role of biophysical interactions within the ijermonding tidal flat sediment dynamics. *Continental Shelf Research* **30**: 1166–1179.

De Baets S, Poesen J, Galindo-Morales P, Knapen A. 2007. Impact of root architecture on the erosion-reducing potential of roots during concentrated flow. *Earth Surface Processes and Landforms* **23**: 1323–1345.

De Baets S, Poesen J, Gyssels G, Knapen A. 2006. Effects of grass roots on the erodibility of topsoils during concentrated flow. *Geomorphology* **76**: 54–67.

De Beer D, Stoodley P, Lewandowski Z. 1996. Liquid flow and mass transport in heterogeneous biofilms. *Water Research* **30**: 2761–2765.

de Deckere EMGT, van de Koppel J, Heip CHR. 2000. The influence of *Corophium volutator* abundance on resuspension. *Hydrobiologia* **426**: 37–42.

Deegan B, Harrington TJ, Dundon P. 2005. Effects of salinity and inundation regime on growth and distribution of *Schoenoplectus triqueter*. *Aquatic Botany* **81**: 199–211

Defew EC, Perkins RG, Paterson DM. 2004. The influence of light and temperature interactions on a natural estuarine microphytobenthic assemblage. *Biofilms* **1**: 21–30.

Deloffre J, Verney R, Lafite R, Lesourd S, Cundy AB. 2007. Sedimentation on intertidal mudflats in the lower part of macrotidal estuaries: Sedimentation rhythms and their preservation. *Marine Geology* **241**: 19–32.

Denny M, Gaylord B. 2002. The mechanics of wave-swept algae. *Journal of Experimental Biology* **205**: 1355–1362.

Denny M, Gaylord B, Helmuth B, Daniel T. 1998. The menace of momentum: dynamic forces on flexible organisms. *Limnology and Oceanography* **43**: 955–968.

Denny MW. 1983. A simple device for recording the maximum force exerted on intertidal organisms. *Limnology and Oceanography* **28**: 1259–1274.

Denny MW. 1988. *Biology and the mechanics of the wave swept environment*. Princeton University Press, Princeton, New Jersey, USA.

Denny MW. 2000. Limits to optimization: fluid dynamics, adhesive strength and the evolution of shape in limpets shells. *The Journal of Experimental Biology* **203**: 2603–2622.

Di Maio JD, Corkum LD. 1997. Patterns of orientation in Unionids as a function of rivers with differing hydrological variability. *Journal of Mollusckan Studies* **63**: 531–539.

Dijkstra JT, Uittenbogaard RE. 2010. Modelling the interaction between flow and highly flexible aquatic vegetation. *Water Resoures Research* **46**: W12547.

Dillon RT. 2000. *The ecology of freshwater molluscs*. Cambridge University Press. Cambridge, UK.

Dittrich A, Schmedtje U. 1995. Indicating shear stress with FST-hemispheres–effects of stream-bottom topography and water depth. *Freshwater Biology* **34**: 107–121.

Dodds WK. 2003. The role of periphyton in phosphorous retention in shallow freshwater aquatic systems. *Journal of Phycology* **39**: 840–849.

Dolmer P. 2000a. Algal concentration profiles above mussel beds. *Journal of Sea Research* **43**: 113–119.

Dolmer P. 2000b. Feeding activity of mussels Mytilus edulis related to near-bed currents and phytoplankton biomass. *Journal of Sea Research* **44**: 221–231.

Donovan DA, Bingham BL, From M, Fleisch AF, Loomis ES. 2003. Effects of barnacle encrustacean on the swimming behaviour, energetics, morphometry, and drag coefficient of the scallop *Chlamys hastata*. *Journal of the Marine Biological Association UK* **83**: 1–7.

Doroudian B, Hurther D, Lemmin U. 2007. Discussion of "Turbulence Measurements with Acoustic Doppler Velocimeters" by Carlos M. García, Mariano I. Cantero, Yarko Niño, and Marcelo H. García. *Journal of Hydraulic Engineering, ASCE* **133**: 1286–1289.

Doyle MW, Stanley EH. 2006. Exploring potential spatial-temporal links between fluvial geomorphology and nutrient-periphyton dynamics in streams using simulation models. *Annals of the Association of American Geographers* **96**: 687–698.

Droppo IG. 2009. Biofilm structure and bed stability of five contrasting freshwater sediments. *Marine and Freshwater Research* **60**: 690–699.

Droppo IG, Jaskot C, Nelson T, Milne J, Charlton M. 2007. Aquaculture waste sediment stability: Implications for waste migration. *Water, Air and Soil Pollutution* **183**: 59–68.

Duarte CM, Terrados J, Agawin NSR, Fortes MD, Bach S, Kenworthy WJ. 1997. Response of a mixed Philippine seagrass meadow to experimental burial. *Marine Ecology Progress Series* **147**: 285–294.

Dubi AM. 1995. *Damping of water waves by submerged vegetation: A case study on Laminaria hyperborea*. PhD Thesis, University of Trondheim, Norway.

Dubi AM, Tørum A. 1996. Wave energy dissipation in kelp vegetation. In: Edge BL (ed) *25th Coastal Engineering Conference. American Society of Civil Engineers*. pp. 2626–2639

Dubois S, Commito JA, Olivier F, Retiere C. 2006. Effects of epibionts on *Sabellaria alveolata* (L.) biogenic reefs and their associated fauna in the Bay of Mont Saint-Michel. *Estuarine, Coastal and Shelf Science* 68: 635–646.

Dufour SC, Desrosiers G, Long B, Lajeunesse P, Gagnoud M, Labrie J, Archambault P, Stora G. 2005. A new method for three-dimensional visualisation and quantification of biogenic structures in aquatic sediments using axial tomodensitometry. *Limnology and Oceanography: Methods* 3: 372–380.

Dukowska M, Sczzerkowska E, Grzybkowska M, Tszydel M, Penczak T. 2007. Effect of flow manipulations on benthic fauna communities in a lowland river: interhabitat comparison. *Polish Journal of Ecology* 55: 101–112.

Dullien FAL. 1979. Porous media, fluid transport and pore structure. San Diego, CA, Academic Press.

Dunn C, Lopez F, García MH. 1996. *Mean flow and turbulence in a laboratory channel with simulated vegetation*. Civil Engineering Studies Report, Hydraulic Engineering Series 51, Hydrosystems Laboratory, Department of Civil Engineering, University of Illinois at Urbana-Champaign, Urbana, IL, USA.

Dupraz C, Reid PR, Braissant O, Decho AW, Norman SR, Visscher PT. 2009. Processes of carbonate precipitation in modern microbial mats. *Earth Science Reviews* 96(3), 141–162.

Dürr S, Thomason JC. 2010. Biofouling. Blackwell Publishing Ltd. Oxford.

Durst F, Melling A, Whitelaw JH. 1981. *Principles and Practice of Laser Doppler Anemometry, Second Edition*. Academic Press, London, UK.

Dworschak PC. 1981. The pumping rates of the burrowing shrimp *Upogebia pusilla* (Petagna) (Decapoda: Thalassinidea). *Journal of Experimental Marine Biology and Ecology* 52: 25–35.

Dworschak PC. 1983. The biology of *Upogebia pusilla* (Petagna) (Decapoda, Thalassinidea). I. The burrows. *Marine Ecology* 4: 19–43.

Dworschak PC. 1987a. Feeding behaviour of *Upogebia pusilla* and *Callianassa tyrrhena* (Crustacea, Decapoda, Thalassinidea). *Investigacion Pesquera* 51 (Supp. 1): 421–429.

Dworschak PC. 1987b. Burrows of *Solecurtus strigilatus* (Linne) and *S. multistriatus* (Scacchi). *Senckenbergiana Marit* 19: 131–147.

Dworschak PC. 1998. The role of tegumental glands in burrow construction by two Mediterranean Callianassid shrimp. *Senckenbergiana Marit* 28: 143–149.

Dworschak PC. 2002. The burrows of *Callianassa candida* (Olvi 1792) and *C.whitei* Sakai 1999 (Crustacea: Decapoda: Thalassinidea). In: Bright M, Dworschak PC, Stchowitsch M. (eds) *The Vienna School of Marine Biology: A Tribute to Jorg Ott*. Facultas Universitatsverlag, Wien, Asutria. pp. 63–71.

Dworschak PC. 2004. Biology of Mediterranean and Caribbean Thalassinidea (Decapoda). In: Tamaki A. (ed) *Proceedings of the Symposium on "Ecology of Large Bioturbators in Tidal Flats and Shallow Sublittoral Sediments* 1–2 Nov 2003, Nagasaki. Nagasaki University, pp. 15–22.

Dworschak PC, Ott JA. 1993. Decapod burrows in mangrove-channel and back-reef environments at the Atlantic Barrier Reef, Belize. *Ichnos* 2: 277–290.

Dworschak PC, Pervesler P. 1988. Burrows of *Callianaasa bouvieri* NOBILI 1904 from Safaga (Egypt, Red Sea) with some remarks on the biology of the species. *Senckenbergiana Marit* 20: 1–17.

Eckman JE. 1983a. Flow perturbation by a protruding animal tube affects rates of sediment microbial colonization. *Eos* 64: 1042.

Eckman JE. 1983b. Hydrodynamic processes affecting benthic recruitment. *Limnology and Oceanography* 28: 241–257.

Eckman JE, Nowell ARM. 1984. Boundary skin friction and sediment transport about an animal-tube mimic. *Sedimentology* 31: 851–862

Eckman JE, Nowell ARM, Jumars PA. 1981. Sediment destabilization by animal tubes. *Journal of Marine Research* 39: 361–374.

Elwany MHS, Flick RE. 1996. Relationship between kelp beds and beach width in Southern California. *Journal of Waterway, Port, Coastal and Ocean Engineering* 122: 34–37.

Elwany MHS, O'Reilly WC, Guza RT, Flick RE. 1995. Effect of Southern California kelp beds on waves. *Journal of Waterway Port Coastal and Ocean Engineering-Asce* 121: 143–150.

Encalada A, Calles J, Ferreira V, Canhoto C, Graça MAS. 2010. Effects of riparian land use on the relationship between benthic communities and leaf litter processing in tropical montane forest streams. *Freshwater Biology* 55: 1719–1733.

Englund VPM, Heino MP. 1996. Valve movement of the freshwater mussel *Andonta anatina*: a reciprocal transplant experiment between lake and river. *Hydrobiologia* 328: 49–56.

Ennos AR. 1990. The anchorage of leek Seedlings: The effect of root length and soil strength. *Annals of Botany* 65: 409–416.

Erduarn KS, Kutija V. 2003. Quasi-3D numerical model for flow through vegetation. *Journal of Hydroinformatics* 3: 189–202.

Ertman SC, Jumars PA. 1988. Effects of bivalve siphonal currents on the settlement of inert particles and larvae. *Journal of Marine Research* 46: 797–813.

Eschweiler N, Buschbaum C. 2011. Alien epibiont (*Crassostrea gigas*) impacts on native periwinkles (*Littorina littorea*). *Aquatic Invasions* 6: 281–190.

Eymann M. 1989. Drag on single larvae of the black fly *Simulium vittatum* (Diptera: SImuliidae) in a thin, growing boundary layer. *Journal of the North American Benthological Society* 7: 109–116.

Fager EW. 1964. Marine sediments: effects of a tube-building polychaete. *Science* 143: 356–359.

Farrow FE. 1971. Back-reef and lagoonal environments of Aldabra Atoll distinguished by crustacean burrows. *Symposia of the Zoological Society of London* 28: 455–500.

Fathi-Moghadam M, Kouwen N. 1997. Non-rigid, non-submerged, vegetative roughness on floodplains. *Journal of Hydraulic Engineering, ASCE* 123: 51–57.

Fauchald K, Jumars PA. 1979. The diet of worms: a study of polychaete feeding guilds. *Oceanography Marine Biology Annual Review* 17: 193–284.

Feagin RA, Irish JL, Möller I, Williams AM, Colón-Rivera RJ, Mousavi ME. 2011. Short communication: Engineering properties of wetland plants with application to wave attenuation. *Coastal Engineering* 58: 251–255.

Feller C, Brown GG, Blanchart E, Deleporte P, Chernyanskii SS. 2003. Charles Darwin, earthworms and the natural sciences: various lessons from past to future. *Agriculture, Ecosystems & Environment* 99: 29–49.

Fernandes S, Sobral P, van Duren L. 2007. Clearance rates of *Cerastoderma edule* under increasing current velocity. *Continental Shelf Research* 27: 1104–1115.

Fernholz H, Janke G, Schober M, Wagner PM, Warnack D. 1996. New developments and applications of skin friction measuring techniques. *Measurement Science and Technology* 7: 1396–1409.

Ferreira V, Gonçalves AL, Godbold DL, Canhoto C. 2010. Effect of increased atmospheric CO_2 on the performance of an aquatic detritivore through changes in water temperature and litter quality. *Global Change Biology* 16: 3284–3296.

Fiala-Médioni A. 1978a. Filter-feeding ethology of benthic invertebrates (Ascidians). III. Recording of water current in situ – rate and rhythm of pumping. *Marine Biology* 45: 185–190.

Fiala-Médioni A. 1978b. Filter-feeding ethology of benthic invertebrates (Ascidians). IV. Pumping rate, filtration, filtration efficiency. *Marine Biology* 48: 243–249.

Finelli CM, Hart DD, Fonseca DM. 1999. Evaluating the spatial resolution of an Acoustic Doppler Velocimeter and the consequences for measuring near-bed flows. *Limnology and Oceanography* **44**: 1793–1801.

Finelli CM, Hart DD, Merz RA. 2002. Stream insects as passive suspension feeders: effects of velocity and food concentration on feeding performance. *Oecologia* **131**: 145–153.

Finley PJ, Khoo CP, Chin JP. 1966. Velocity measurements in a thin turbulent water layer. *La Houille Blanche* **21**: 713–721.

Finnigan J. 2000. Turbulence in plant canopies. *Annual Review of Fluid Mechanics* **32**: 519–571.

Flemming HC. 2011. "The Perfect Slime", Colloids and Surfaces B: Biointerfaces, **86**: 251–259.

Flemming HC, Neu TR, Wozniak DJ. 2007. The EPS matrix: the "House of Biofilm Cells." *Journal of Bacteriology* **189**: 7945–7947.

Flemming HC, Wingender J. 2001. Relevance of microbial extracellular polymeric substances (EPSs) – Part I: structural and ecological aspects. *Water Science and Technology* **43**: 1–8.

Flynn KM, McKee KL, Mendelssohn IA. 1995. Recovery of freshwater marsh vegetation after a saltwater intrusion event. *Oecologia* **103**: 63–72.

Folkard AM. 2005. Hydrodynamics of model *Posidonia oceanica* patches in shallow water. *Limnology and Oceanography* **50**(5): 1592–1600.

Folkard AM, Gascoigne JC. 2009. Hydrodynamics of discontinuous mussel beds: Laboratory flume simulations. *Journal of Sea Research* **62**: 250–257.

Fonseca MS, Cahalan JA. 1992. A preliminary evaluation of wave attenuation by four species of seagrass. *Estuarine, Coastal and Shelf Science* **35**: 565–576.

Fonseca MS, Fisher JS, Zieman JC, Thayler GW. 1982. Influence of the seagrass, *Zostera marina* L., on current flow. *Estuarine Coastal and Shelf Science* **15**: 351–364.

Fonseca MS, Fisher JS. 1986. A comparison of canopy friction and sediment movement between four species of seagrass with reference to their ecology and restoration. *Marine Ecology Progress Series* **29**: 15–22.

Fonseca MS, Koehl MAR, Kopp BS. 2007. Biomechanical factors contributing to self-organization in seagrass landscapes. *Journal of Experimental Marine Biology and Ecology* **340**: 227–246.

Fonseca MS, Koehl MAR. 2006. Flow in seagrass canopies: The influence of patch width. *Estuarine, Coastal and Shelf Science* **67**: 1–9.

Forster S, Zettler ML. 2004. The capacity of the filter-feeding bivalve *Mya aernaria* L. to affect water transport in sandy beds. *Marine Ecology* **144**: 1183–1189.

Franca M, Lemmin U. 2006. Eliminating velocity aliasing in acoustic Doppler velocity profiler data. *Measurement Science and Technology* **17**: 313–322.

Francoeur SN, Biggs BJF. 2006. Short-term effects of elevated velocity and sediment abrasion on benthic algal communities. *Hydrobiologia* **561**: 59–69.

Francois F, Gerino M, Stora G, Durbec J-P, Poggiale J-C. 2002. Functional approach to sediment reworking by gallery-forming macrobenthic organisms: modelling and application with the polychaete *Nereis diversicolor*. *Marine Ecology Progress Series* **229**: 127–136.

Frank DM, Ward JE, Shumway SE, Holohan BA, Gray C. 2008. Application of particle image velocimetry to the study of suspension feeding in marine invertebrates. *Marine and Freshwater Behaviour and Physiology* **41**: 1–18.

Fréchette M, Butman CA, Geyer WR. 1989. The importance of boundary-layer flows in supplying phytoplankton to the benthic suspension feeder, *Mytilus edulis* L. *Limnology and Oceanography* **34**: 19–36.

Freeman GE, Rahmeyer WH, Copeland RR. 2000. Determination of resistance due to shrubs and woody vegetation. *ERDC/CHL Technical Report* **00-25**, US Army Corps of Engineers, Vicksburg, Mississippi, USA.

French CE, French JR, Clifford NJ, Watson CJ. 2000. Sedimentation-erosion dynamics of abandoned reclamations: the role of waves and tides. *Continental Shelf Research* 20: 1711–1733.

French JR, Spencer T, Murray AL, Arnold NS. 1995. Geostatistical analysis of sediment deposition in two small tidal wetlands, Norfolk, U.K. *Journal of Coastal Research* 11: 308–321.

Friedrichs M, Graf G. 2009. Characteristic flow patterns generated by macrozoobenthic structures. *Journal of Marine Systems* 75: 348–359.

Friedrichs M, Graf G, Springer B. 2000. Skimming flow induced over a simulated polychaete tube lawn at low population densities. *Marine Ecology Progress Series* 192: 219–228.

Friedrichs M, Leipe T, Peine F, Graf G. 2009. Impact of macrozoobenthic structures on near-bed sediment fluxes. *Journal of Marine Systems* 75: 336–347.

Frostick LE, McLelland SJ, Mercer TG. 2011. *Users guide to Physical Modelling and Experimentation: Experience of the HYDRALAB Network*. IAHR Design Manual, CRC Press/ Balkema, Leiden, The Netherlands.

Frutiger A, Schib JL. 1993. Limitations of FST hemispheres in lotic benthos research. *Freshwater Biology* 30: 463–474.

Fukuhara H. 1987. The effect of tubificids and chironomids on particle redistribution of lake sediment. *Ecological Research* 2: 255–264.

Fuller RL, Doyle S, Levy L, Owens J, Shope E, Vo L, Wolyniak E, Small MJ, Doyle MW. 2011. Impact of regulated releases on periphyton and macroinvertebrate communities: the dynamic relationship between hydrology and geomorphology in frequently flooded rivers. *River Research and Applications* 27: 630–645.

Furukawa K, Wolanski E, Mueller H. 1997. Currents and sediment transport in mangrove forests. *Estuarine, Coastal and Shelf Science* 44: 301–310.

Gacia E. 2001. Sediment retention by a Mediterranean *Posidonia oceanica* meadow: The balance between deposition and resuspension. *Estuarine, Coastal and Shelf Science* 52: 505–514.

Gacia E, Duarte CM. 2001. Sediment retention by a Mediterranean *Posidonia oceanica* meadow: The balance between deposition and resuspension. *Estuarine, Coastal and Shelf Science* 52: 505–514.

Gacia E, Granata TC, Duarte CM. 1999. An approach to measurement of particle flux and sediment retention within seagrass (*Posidonia oceanica*) meadows. *Aquatic Botany* 65: 255–268.

Gainswin BE, House WA, Leadbeater BSC, Armitage PD. 2006a. Kinetics of phosphorus release from a natural mixed grain-size sediment with associated algal biofilms. *Science of the Total Environment* 360: 127–141.

Gainswin BE, House WA, Leadbeater BSC, Armitage PD, Patten J. 2006b. The effects of sediment size fraction and associated algal biofilms on the kinetics of phosphorus release. *Science of the Total Environment* 360: 142–157.

Gambi MC, Nowell ARM, Jumars PA. 1990. Flume observations on flow dynamics in *Zostera marina* (eelgrass) beds. *Marine Ecology Progress Series* 61: 159–169.

García CM, Cantero MI, Niño Y, García MH. 2005. Turbulence measurements with acoustic Doppler velocimeters. *Journal of Hydraulic Engineering* 131: 1062–1073.

García CM, Cantero MI, Niño Y, García MH. 2007. Closure to "Turbulence Measurements with Acoustic Doppler Velocimeters" by Carlos M. García, Mariano I. Cantero, Yarko Niño, and Marcelo H. García. *Journal of Hydraulic Engineering, ASCE* 133: 1289–1292.

García CM, García MH. 2006. Characterization of flow turbulence in large-scale bubble-plume experiments. *Experiments in Fluids* 41: 91–101.

García CM, Jackson PR, García MH. 2006. Confidence intervals in the determination of turbulence parameters. *Experiments in Fluids* 40: 514–522.

Garcia X-F, Ricardo AM, Blanckaert K. 2011. The role of turbulent coherent structures in invertebrate drift. In: Venditti JG, Best JL, Church MA, Hardy RJ. (eds) *Proceedings of CFSII: Coherent Flow Structures in Geophysical Flows at Earth's Surface, Burnaby, Canada, August 3–5 2011*. Simon Fraser University, Burnaby, Canada.

García-March JR, García-Carrascosa AM, Pena Cantero, Wang Y-G. 2007b. Population structure, mortality and growth of *Pinna nobilis* Linnaeus, 1758 (Mollusca, Bivalvia) at different depths in Moraira bay (Alicante, Western Mediterranean). *Marine Biology* 150: 861–871.

Gardner LR, Sharma P, Moore WS. 1987. A regeneration model for the effect of bioturbation by fiddler crabs on ^{210}Pb profiles in salt marsh sediments. *Journal of Environmental Radioactivity* 5: 25–36.

Gascoigne JC, Beadman HA, Saurel C, Kaiser MJ. 2005. Density dependence, spatial scale and patterning in sessile biota. *Oecologia* 145: 371–381.

Gaylord B, Denny MW, Koehl MAR. 2003. Modulation of wave forces on kelp canopies by alongshore currents. *Limnology and Oceanography* 48: 860–871.

Gerdol V, Hughes RG. 2008. Effect of *Corophium volutator* on the abundance of benthic diatoms, bacteria and sediment stability in two estuaries in southeastern England. *Marine Ecology Progress Series* 114: 109–115.

Gerino M. 1990. The effects of bioturbation on particle redistribution in Mediterranean coastal sediment. Preliminary results. *Hydrobiologia* 207: 251–258.

Gerino M, Stora G, Francois F, Gilbert JC, Poggiale JC, Mermillod-Blondin F, Desrosier G, Vervier P. 2003. Macro-invertebrate functional groups in freshwater and marine sediments: a common mechanistic classification. *Vie Milieu* 53: 222–231.

Ghisalberti M, Nepf HM. 2002. Mixing layers and coherent structures in vegetated aquatic flows. *Journal of Geophysical Research* 107(C2), DOI:10.1029/2001 JC000871.

Ghisalberti M, Nepf HM. 2006. The structure of the shear layer over rigid and flexible canopies. *Environmental Fluid Mechanics* 6: 277–301.

Gilbert F, Hulth S, Grossi V, Poggiale JC, Desrosiers G, Rosenberg R, Gerino M, Francois-Carcaillet F, Michaud E, Stora G. 2007. Sediment reworking by marine benthic species from the Gullmar Fjord (Western Sweden): importance of faunal biovolume. *Journal of Experimental Marine Biology and Ecology* 348: 133–144.

Giovannetti E, Montefalcone M, Morri C, Bianchi CN, Albertelli G. 2010. Early warning response of *Posidonia oceanica* epiphyte community to environmental alterations (Ligurian Sea, NW Mediterranean). *Marine Pollution Bulletin* 60: 1031–1039.

Godillot R, Caussade B, Ameziane T, Capblanq J. 2001. Interplay between turbulence and periphyton in rough open-channel flow. *Journal of Hydraulic Research* 39: 227–239.

Goessmann C, Hemelrijk C, Huber R. 2000. The formation and maintenance of crayfish hierarchies: behavioural and self-structuring properties. *Behavioral Ecology and Sociobiology* 48: 418–428.

Goring DG, Nikora VI. 2002. Despiking acoustic Doppler velocimeter data. *Journal of Hydraulic Engineering, ASCE* 128: 117–126.

Gouleau D, Jouanneau JM, Weber O, Sauriau PG. 2000. Short- and long-term sedimentation on Montportail-Brouage intertidal mudflat, Marennes-Oleron Bay (France). *Continental Shelf Research* 20: 12–13.

Gourlay MR. 1970. Discussion of: Flow resistance in vegetated channels, by Kouwen N, Unny TE and Hill HM. *Journal of the Irrigation and Drainage Division of the American Society of Civil Engineers* 96: 351–357.

Graba M, Moulin FY, Boulêtreau S, Garabétian F, Kettab A, Eiff O, Sánchez Pérez J-M, Sauvage S. 2010. Effect of near bed turbulence on chronic detachment of epilithic biofilm: Experimental and modelling approaches. *Water Resources Research* 46: W11531.

Grabowski RC, Droppo IG, Wharton G. 2011. Erodibility of cohesive sediment: The importance of sediment properties. *Earth-Science Reviews* 105(3–4): pp. 101–120.

Graf WH, Istiarto I. 2002. Flow pattern in the scour hole around a cylinder. *Journal of Hydraulic Research* 40: 13–20.

Graf WH, Yulistiyanto B. 1998. Experiments on flow around a cylinder; the velocity and vorticity fields. *Journal of Hydraulic Research* 36: 637–654.

Graham AA. 1990. Siltation of stone-surface periphyton in rivers by clay-sized particles from low concentrations in suspension. *Hydrobiologia* 199: 107–115.

Graham GW, Manning AJ. 2007. Floc size and settling velocity within a *Spartina anglica* canopy. *Continental Shelf Research* 27: 1060–1079.

Gran K, Paola C. 2001. Riparian vegetation controls on braided stream dynamics. *Water Resources Research* 37: 3275–3283.

Grassle JP, Snelgrove PVR, Butman CA. 1992. Larval habitat choice in still water and flume flows by the opportunistic bivalve *Mulinia lateralis*. *Netherlands Journal of Sea Research* 30: 33–44.

Green JC. 2005. Modelling flow resistance in vegetated streams: review and development of new theory. *Hydrological Processes* 19: 1245–1259.

Green JC, Richards KS. 1998. Investigation into the validity of resistance-discharge relationships for trailing aquatic plants. In: Wheater H, Kirby C. (ed) *Hydrology in a Changing Environment*. John Wiley, Chichester, UK, pp. 413–420.

Green MO, Hewitt JE, Thrush SF. 1998. Seabed drag coefficient over natural beds of horse mussels (*Atrina zelandica*). *Journal of Marine Research* 56: 613–637.

Green S, Visser AW, Titelman J, Kiorboe T. 2003. Escape responses of copepod *nauplii* in the flow field of the blue mussel, *Mytilus edulis*. *Marine Biology* 142: 727–733.

Greenway DR. 1987. Vegetation and slope stability. In: M.G. Anderson and K.S. Richards, eds. Slope Stability. John Wiley & Sons Ltd., Chichester, UK, pp. 187–230.

Griffis RB, Chavez FL. 1988. Effects of sediment type on burrows of *Callianassa californiensis* Dana and *C. gigas* Dana. *Journal of Experimental Marine Biology and Ecology* 117: 239–253.

Griffis RB, Suchanek TH. 1991. A model of burrow architecture and trophic modes in thalassinidean shrimp (Decapoda: Thalassinidea). *Marine Ecology Progress Series* 79: 171–183.

Grizzle RE, Langan R, Howell WH. 1992. Growth responses of suspension-feeding bivalve molluscs to changes in water flow: differences between siphonate and nonsiphonate taxa. *Journal of Experimental Marine Biology and Ecology* 162: 213–228.

Guan R-Z. 1994. Burrowing behaviour of signal crayfish, *Pacifastacus leniusculus* (Dana), in the River Great Ouse, England. *Freshwater Forum* 4: 155–168.

Gurnell AM, Bertoldi W, Corenblit D. 2012. Changing river channels: The roles of hydrological processes, plants and pioneer fluvial landforms in humid temperate, mixed load, gravel bed rivers. *Earth-Science Reviews* 111: 129–141.

Gurnell AM, Morrissey IP, Boitsidis AJ, Bark T, Clifford NJ, Petts GE, Thompson K. 2006. Initial adjustments within a new river channel: Interactions between fluvial processes, colonizing vegetation, and bank profile development. *Environmental Management* 38: 580–596.

Gurnell AM, van Oosterhout MP, de Vlieger B, Goodson JM. 2006. Reach-scale interactions between aquatic plants and physical habitat: River Frome, Dorset. *River Research and Applications* 22(6), 667–680.

Gust G. 1988. Skin friction probes for field applications. *Journal of Geophysical Research* 93: 14121–14132.

Gutiérrez JL, Iribarne O. 1999. Role of Holocene beds of the stout razor clam *Tagelus plebeius* in structuring present benthic communities. *Marine Ecology Progress Series* 185: 213–228.

Gutiérrez JL, Jones CG, Strayer DL, Iribarne O. 2003. Mollusks as ecosystem engineers: the role of shell production in aquatic habitats. *Oikos* 101: 79–90.

Gyssels G, Poesen, J. 2003. The importance of plant root characteristics in controlling concentrated flow erosion rates. *Earth Surface Processes and Landforms* 28: 371–384.

Gyssels G, Poesen J, Bochet E, Li Y. 2005. Impact of plant roots on the resistance of soils to erosion by water: a review. *Progress in Physical Geography* 29: 189–217.

Halan B, Buehler K, Schmid A. 2012. Biofilms as living catalysts in continuous chemical syntheses. *Trends in Biotechnology* 30(9): 454–465.

Hancke K, Glud RN. 2004. Temperature effects on respiration and photosynthesis in three diatom-dominated benthic communities. *Aquatic Microbial Ecology* 37: 265–281.

Hanratty TJ. 1991. Use of the polarographic method to measure wall shear stress. *Journal of Applied Electrochemistry* 21: 1038–1046.

Hanratty TJ, Campbell JA. 1983. Measurement of wall shear stress. In: Goldstein RJ. (ed) *Fluid Mechanics Measurements*. Hemisphere, Washington, DC, USA. pp. 559–615.

Harada K, Latief H, Imamura F. 2000. Study on the mangrove control forest ot reduce tsunami impact. *Proceedings of 12th congress of the IAHR-APD*, Bangkok, Thailand.

Haritonidis JH. 1989. The measurement of wall shear stress. In: Gad-el-Hak M. (ed) *Advances in Fluid Mechanics Measurements, Lecture Notes in Engineering* 45, Springer-Verlag, New York, New York, USA. pp. 229–261.

Harvey GL, Moorhouse TP, Clifford NJ, Henshaw AJ, Johnson MF, MacDonald DW, Reid I, Rice SP. 2011. Evaluating the role of invasive aquatic species as drivers of fine sediment-related river management problems: the case of the signal crayfish (*Pacifastacus leniusculus*). *Progress in Physical Geography* 35: 517–533.

Haslam SM. 1978. *River Plants: The Macrophytic Vegetation of Watercources*. Cambridge University Press, Cambridge, UK.

Hassan MA, Gottesfeld AS, Montgomery DR, Tunnicliffe JF, Clarke GKC, Wynn G, Jones-Cox H, Poirier R, MacIsaac E, Herunter H, McDonald SJ. 2008. Salmon driven bedload transport and bed morphology in mountain streams. *Geophysical Research Letters* 35: L04405.

Haven DS, Morales-Alamo R. 1996. Aspects of biodeposition by oysters and other invertebrate filter-feeders. *Limnology and Oceanography* 11: 487–498.

Heinmayer O, Digialleonardo J, Qian L, Roseijadi G. 2008. Stress tolerance of a subtropical *Crassostrea virginica* population to the combined effects of temperature and salinity. *Estuarine, Coastal and Shelf Science* 79: 179–185.

Hempel C. 1957. Über den Rohrenbau und die Nahrungsaufnahme einiger spioniden der deutschen Küsten. *Helgolander Wissenschaftliche Meeresuntersuchungen* 6: 100–134.

Hendriks IE. Sintes T, Bouma TJ, Duarte CM. 2008. Experimental assessment and modelling evaluation of the effects of the seagrass *Posidonia oceanica* on flow and particle trapping. *Marine Ecology-Progress Series* 356: 163–173.

Hendriks IE, van Duren LA, Herman PMJ. 2006. Turbulence levels in a flume compared to the field: Implications for larval settlement studies. *Journal of Sea Research* 55: 15–29.

Hicks DM, Duncan MJ, Land SN, Tal M, Westaway R. 2008. Contemporary morphological change in braided gravel-bed rivers: new developments from field and laboratory studies, with particular reference to the influence of riparian vegetation. In: Habersack H, Piegay H, Rinaldi M (eds) *Gravel-bed Rivers VI: From Process Understanding to River Restoration*. Elsevier, Amsterdam, The Netherlands. pp. 557–584.

Hill BJ, Allanson BR. 1971. Temperature tolerance of the estuarine prawn *Upogebia africana* (Anomura, Crustacea). *Marine Biology* 11: 337–343.

Hinze JO. 1975. *Turbulence: An Introduction to its Mechanism and Theory, 2nd Edition*. McGraw-Hill, New York, New York, USA.

Hondzo M, Wang H. 2002. Effects of turbulence on growth and metabolism of periphyton in a laboratory flume. *Water Resources Research* 38: 1277.

Hopkinson L, Wynn T. 2009. Vegetation impacts on near bank flow. *Ecohydrology*, online, pp. 1–15.

Horner RR, Welch EB, Seeley MR, Jacoby JM. 1990. Responses of periphyton to changes in current velocity, suspended sediment and phosphorus concentration. *Freshwater Biology* 24: 215–232.

Howard RJ, Mendelssohn IA. 1999. Salinity as a constraint on growth of oligohaline marsh macrophytes. I. Species variation in stress tolerance. *American Journal of Botany* 86: 785–794.

Huai WX, Zeng YH, Xu ZG, Yang ZH. 2009. Three-layer model for vertical velocity distribution in open channel flow with submerged rigid vegetation. *Advances in Water Resources* 32: 487–492.

Huettel M, Gust G. 1992. Impact of bioroughness on interfacial solute exchange in permeable sediments. *Marine Ecology Progress Series* 89: 253–267.

Huettel M, Rusch A. 2000. Transport and degradation of phytoplankton in permeable sediment. *Limnology and Ocenography* 3: 534–549.

Huettel M, Ziebis W, Forster S. 1996. Flow-induced uptake of particulate matter in permeable sediments. *Limnology and Ocenography* 41: 309–322.

Hughes DJ, Atkinson RJA. 1997. A towed video survey of megafaunal bioturbation in the North-Eastern Irish Sea. *Journal of the Marine Biological Association UK* 77: 635–653.

Hultmark M, Leftwich M, Smits AJ. 2007. Flowfield measurements in the wake of a robotic lamprey. *Experiments in Fluids* 43: 683–690.

Hunt HL, Scheibling RE. 2001. Predicting wave dislodgment of mussels: variation in attachment strength with body size, habitat, and season. *Marine Ecology Progress Series* 213: 157–164.

Hunt MJ, Alexander CG. 1991. Feeding mechanisms in the barnacle *Tetraclita squamosa* (Bruguiére). *Journal of Experimental Marine Biology and Ecology* 154: 1–28.

Hunter TN, Peakall J, Biggs SR. 2011. Ultrasonic velocimetry for the *in situ* characterisation of particulate settling and sedimentation. *Minerals Engineering* 24: 416–423.

Hurther D, Lemmin U. 1998. A constant-beam-width transducer for 3D Doppler profile measurements in open-channel flows. *Measurement Science and Technology* 9: 1706–1714.

Hurther D, Lemmin U. 2001. A correction method for turbulence measurements with a 3-D Acoustic Doppler Velocity Profiler. *Journal of Atmospheric and Oceanic Technology* 18: 446–458.

Husrin S, Oumeraci H. 2009. Parameterization of coastal forest vegetation and hydraulic resistance coefficients for tsunami modelling, 4th Annual International Workshop and Expo on Sumatra Tsunami Disaster and Recovery.

Idestam-Almquist J, Kautsky L. 1995. Plastic responses in morphology of *Potamogeton pectinatus* L. to sediment and above sediment conditions at two sites in the northern Baltic proper. *Aquatic Botany* 52: 205–216.

Ikeda S, Kanazawa M. 1996. Three-dimensional organized vortices above flexible plants. *Journal of Hydraulic Engineering* 122: 634–640.

Ikeda S, Yamada T, Toda Y. 2001. Numerical study on turbulent flow and honami in and above flexible plant canopy. *International Journal of Heat and Fluid Flow* 22: 252–258.

Iribarne O, Bortolus A, Botto F. 1997. Between-habitat differences in burrow characteristics and trophic modes in the southwestern Atlantic burrowing crab *Chasmagnathus granulata*. *Marine Ecology Progress Series* 155: 132–145.

Iribarne O, Botto F. 1998. Orientation of the extant stout razor clam *Tagelus plebeius* in relation to current direction: its palaeoecologic implications. *Journal of Shellfish Research* 17: 165–168.

Jackson CR, Churchill PF, Roden EE. 2001. Successional changes in bacterial assemblage structure during epilithic biofilm development. *Ecology* 82: 555–566.

Jakob C, Robinson CT, Uehlinger U. 2003. Longitudinal effects of experimental floods on stream benthos downstream from a large dam. *Aquatic Science* 65: 223–231.

James CS, Goldbeck UK, Patini, A, Jordanova AA. 2008. Influence of foliage on flow resistance of emergent vegetation. *Journal of Hydraulic Research* 46: 536–542.

James WF, Barko JW. 2000. *Sediment resuspension dynamics in canopy- and meadow-forming submersed macrophyte communities*. Aquatic Plant Control Research Program, U.S. Army Corps of Engineers, USA.

Järvelä J. 2002. Flow resistance of flexible and stiff vegetation: a flume study with natural plants. *Journal of Hydrology* 269: 44–54.

Järvelä, J. 2004. Determination of flow resistance caused by non-submerged woody vegetation. *International Journal of River Basin Management* 2: 61–70.

Järvelä J. 2005. Effect of submerged flexible vegetation on flow structure and resistance. *Journal of Hydrology* 307: 233–241.

Johnson MF, Reid I, Rice SP, Wood PJ. 2009. Stabilisation of fine gravels by net-spinning caddisfly larvae. *Earth Surface Processes and Landforms* 34: 413–423.

Johnson MF, Rice SP, Reid I. 2010. Topographic disturbance of subaqueous gravel substrates by signal crayfish (*Pacifastacus leniusculus*). *Geomorphology* 123: 260–278.

Johnson MF, Rice SP, Reid I. 2011. Increase in coarse sediment transport associated with disturbance of gravel river beds by signal crayfish (*Pacifastacus leniusculus*). *Earth Surface Processes and Landforms* 36: 1680–1692.

Jones HFE, Pilditch CA, Bryan KR, Hamilton DP. 2011. Effects of infaunal bivalve density and flow speed on clearence rates and near-bed hydrodynamics. *Journal of Experimental Marine Biology and Ecology* 40: 20–28.

Jones, J., Collins, A., Naden, P., Sear, D., 2012. The relationship between fine sediment and macrophytes in rivers. *River Res. Applic.* 28(7), 1006–1018. doi:10.1002/rra.1486.

Jordanova AA, James CS. 2003. Experimental study of bedload transport through emergent vegetation. *Journal of Hydraulic Engineering* 129(6), 474–478.

Jørgensen BB, Glud RN, Holby O. 2005. Oxygen distribution and bioirrigation in Arctic fjord sediments (Svalbard, Barents Sea). *Marine Ecology Progress Series* 292: 85–95.

Jørgensen CB. 1990. *Bivalve filter feeding: hydrodynamics, bioenergetics, physiology and ecology*. Olsen and Olsen, Fredensborg, Denmark.

Jørgensen CB, Famme P, Kristensen HS, Larsen PS, Møhlenberg F, Riisgård HU. 1986. The bivalve pump. *Marine Ecology Progress Series* 34: 69–77.

Kadlec RH. 1990. Overland flows in wetlands: vegetation resistance. *Journal of Hydraulic Engineering* 116: 691–706.

Kao DTY, Barfield BJ. 1978. Prediction of flow hydraulics for vegetated channels. *Transactions of the ASAE*. Paper No. 76–2057: 489–495.

Karathanasis AD, Potter CL, Coyne MS. 2003. Vegetation effects on fecal bacteria, BOD, and suspended solid removal in constructed wetlands treating domestic wastewater. *Ecological Engineering* 20: 157–169.

Kat PW. 1982. Effects of population density and substratum type on growth and migration of *Elliptio complanata* (Bivalvia: Unionidae). *Malacological Review* 15: 119–127.

Katrak G, Dittmann S, Seuront L. 2008. Spatial variation in burrow morphology of the mud shore crab *Helograpsus haswellianus* (Brachyura, Grapsidae) in South Australian salt-marshes. *Marine and Freshwater Research* 59: 902–911.

Katz LC. 1980. Effects of burrowing by the fiddler crab, *Uca pugnax* (Smith). *Estuarine, Coastal and Shelf Science* 11: 233–237.

Keane RD, Adrian RJ. 1992. Theory of cross-correlation analysis of PIV images. *Applied Scientific Research* 49: 191–215.

Keddy PA. 2000. *Wetland Ecology: Principles and Conservation*. Cambridge University Press, Cambridge, UK.

Kemp J, Harper D, Crosa G. 2000. The habitat-scale ecohydraulics of rivers. *Ecological Engineering* **16**: 17–29.

Kemp P, Bertness MD. 1984. Snail shape and growth rates: evidence for plastic shell allometry in *Littorina littorea. Proceedings of the National Academy Science* **81**: 811–813.

Kempf G.1937. On the effect of roughness on the resistance of ships. *Transactions of the Institution of Naval Architects* **79**: 109–119

Kershaw PJ, Swift DJ, Pentreath RJ. Lovett MB. 1984. The incorporation of plutonium, americium and curium into the Irish Sea seabed by biological activity. *Science of the Total Environment* **40**: 61–81.

Keylock CJ, Hardy RJ, Parsons DR, Ferguson RI, Lane SN, Richards KS. 2005. The theoretical foundations and potential for large-eddy simulation (LES) in fluvial geomorphic and sedimentological research. *Earth-Science Reviews* **71**: 271–304.

Kidwell SM. 1986. Taphonomic feedback in Miocene assemblages: testing the role of dead hardparts in benthic communities. *Palaios* **1**: 239–255.

Kim BK, Jackman AP, Triska FJ. 1990. Modeling transient storage and nitrate uptake kinetics in a flume containing a natural periphyton community. *Water Resources Research* **26**: 505–515.

Kim BKA, Jackman AP, Triska FJ. 1992. Modelling biotic uptake by periphyton and transient hyporheic storage of nitrate in a natural stream. *Water Resources Research* **28**: 2743–2752.

Kingston MB, Gough JS. 2009. Vertical migration of a mixed-species euglena (Euglenophyta) assemblage inhabiting the high-intertidal sands of Nye beach, Oregon. *Journal of Phycology* **45**: 1021–1029.

Kirtley DW, Tanner WF. 1968. Sabellariid worms; builders of a major reef type. *Journal of Sedimentary Research* **38**: 73–78.

Klopstra D, Barneveld HJ, Van Noortwijk JM, Van Velzen EH. 1997. Analytical model for hydraulic roughness of submerged vegetation, 27th international IAHR conference, pp. 775–780.

Knutson PL. 1988. Role of coastal marshes in energy dissipation and shore protection. In: Hook DD. (ed) *The Ecology and Management of Wetlands.* I. Croom Helm, London, UK. pp. 161–175.

Knutson PL, Seeling WN, Inskeep MR. 1982. Wave damping in *Spartina alterniflora* marshes. *Wetlands* **2**: 87–104.

Kobayashi N, Raichle AW, Asano T. 1993. Wave attenuation by vegetation. *Journal of Waterway, Port, Coastal, and Ocean Engineering* **119**: 30–48.

Koch EW. 1999. Sediment resuspension in a shallow *Thalassia testudinum* banks ex Konig bed. *Aquatic Botany* **65**: 269–280.

Koch EW. 2001. Beyond light: physical, geological, and geochemical parameters as possible submerded aquatic vegetation habitat requirements. *Estuaries* **24**: 1–17.

Koch EW. 2002. Impact of boat-generated waves on seagrass habitat. *Journal of Coastal Research* **37**: 66–74.

Koch EW, Barbier EB, Silliman BR, Reed DJ, Perillo GME, Hacker SD. Granek EF, Primavera JH, Muthiga N, Polansky S, Halpern BS, Kennedy CJ, Kappel CV, Wolanksi E. 2009. Non-linearity in ecosystem services: temporal and spatial variability in coastal protection. *Frontiers in Ecology and the Environment* **7**: 29–37.

Koch EW, Gust G. 1999. Water flow in tide- and wave-dominated beds of the seagrass *Thalassia testudinum. Marine Ecology Progress Series* **184**: 63–72.

Koch EW, Sanford LP, Chen S-N, Shafer DJ, McKee Smith J. 2006. *Waves in seagrass systems: review and technical recommendations.* US Army Corps of Engineers U.S. Army Engineer Research and Development Center, USA.

Koehl MAR. 1977. Effects of sea anemones on the flow forces they encounter. *Journal of Experimental Biology* **69**: 87–105.

Koehl MAR. 1996. When does morphology matter? *Annual Review of Ecology and Systematics* **27**: 501–542.

Kolmogorov AN. 1941. The local structure of turbulence in incompressible viscous fluid for very large Reynolds numbers. First published in Russian in *Doklady Akademii Nauk SSSR* **30**(4). Translation by V Levin reprinted in 1991 in *Proceedings of the Royal Society of London A* **434**(1890): 9–13. See also corrections printed in 1994 in *Journal of Fluid Mechanics* **295**: 406–408.

Konhauser, K. 2007. Introduction *to Geomicrobiology*. Blackwell Publishing, Oxford.

Kouwen N, Li RM. 1980. Biomechanics of vegetative channel linings. *Journal of the Hydraulics Division of the American Society of Civil Engineers* **106**: 1085–1103.

Kraus NC, Lohrmann A, Cabrera R. 1994. New acoustic meter for measuring 3D laboratory flows. *Journal of Hydraulic Engineering, ASCE* **120**: 406–412.

Kristensen K, Hansen K. 1999. Transport of carbon dioxide and ammonium in bioturbated (*Nereis diversicolor*) coastal, marine sediments. *Biogeochemistry* **45**: 147–168.

Kutija V, Hong HTM. 1996. A numerical model for assessing the additional resistance to flow introduced by flexible vegetation. *Journal of Hydraulic Research* **34**: 99–114.

LaBarbera M. 1981. Water flow patterns in and around three species of articulate brachiopods. *Journal of Experimental Marine Biology and Ecology* **55**: 185–206.

Labiod C, Godillot R, Caussade B. 2007. The relationship between stream periphyton dynamics and near-bed turbulence in rough open-channel flow, *Ecological Modelling* **209**: 78–96.

Lacoul P, Freedman B. 2006. Environmental influences on aquatic plants in freshwater ecosystems. *Environmental Reviews* **14**: 89–136.

Laima M, Brossard D, Sauriau P-G, Girard M, Richard P, Gouleau D, Joassard L. 2002. The influence of long emersion on biota, ammonium fluxes and nitrification in intertidal sediments of Marennes-Oléron Bay, France. *Marine Environmental Research* **53**: 381–402.

Lamouroux N, Mérigoux S, Capra H, Dolédec S, Jowett IG, Statzner B. 2010. The generality of abundance-environment relationships in microhabitats: a comment on Lancaster and Downes (2009). *River Research and Applications* **26**: 915–920.

Lancaster J, Buffin-Bélanger T, Reid I, Rice S. 2006. Flow- and substratum-mediated movement by a stream insect. *Freshwater Biology* **51**: 1053–1069.

Lancaster J, Downes BJ. 2010a. Linking the hydraulic world of individual organisms to ecological processes: putting ecology into ecohydraulics. *River Research and Applications* **26**: 385–403.

Lancaster J, Downes BJ. 2010b. Ecohydraulics needs to embrace ecology and sound science, and to avoid mathematical artefacts. *River Research and Applications* **26**: 921–929.

Larned ST, Nikora VI, Biggs BJF. 2004. Mass-transfer–limited nitrogen and phosphorus uptake by stream periphyton: A conceptual model and experimental evidence. *Limnology and Oceanography* **49**: 1992–2000.

Larsen LG, Harvey JW, Crimaldi JP. 2009. Predicting bed shear stress and its role in sediment dynamics and restoration potential of the Everglades and other vegetated flow systems. *Ecological Engineering* **35**: 1773–1785.

Larsen PS, Riisgård HU. 1997. Biomixing generated by benthic filter feeders: a diffusion model for near-bottom phytoplankton depletion. *Journal of Sea Research* **37**: 81–90.

Larson M. 1995. Model for decay of random waves in surf zone. *Journal of Waterway, Port, Coastal and Ocean Engineering* **121**: 1–12.

Lassen J, Kortegard M, Riisgård HU, Friedrich M, Graf G, Larsen PS. 2006. Down-mixing of phytoplankton above filter-feeding mussels – interplay between water flow and biomixing. *Marine Ecology Progress Series* **314**: 77–88.

Le Hir P, Monbet Y, Orvain F. 2007. Sediment erodability in sediment transport modeling: can we account for biota effects? Continental Shelf Research **27**: 1116–1142.

Lee SY, Fong CW, Wu RSS. 2001. The effects of seagrass (*Zostera japonica*) canopy structure on associated fauna: a study using artificial seagrass units and sampling of natural beds. *Journal of Experimental Marine Biology and Ecology* **259**: 23–50.

Lefebvre A, Thompson CEL, Amos CL. 2010. Influence of *Zostera marina* canopies on uni-directional flow, hydraulic roughness and sediment movement. *Continental Shelf Research* **30**: 1783–1794.

Lehmann J, Wippich MGE. 1995. Oyster attachment and scar preservation of a Late Maastrich-tian ammonite *Hoploscaphites constrictus*. *Acta Palaeontologica Polonica* **40**: 437–440.

Lelieveld SD, Pilditch CA, Green, MO. 2004. Effects of deposit-feeding bivalve (*Macomona liliana*) density on intertidal sediment stability. *New Zealand Journal of Marine and Fresh-water Research* **38**: 115–128.

Lemmin U, Rolland T. 1997. Acoustic Velocity Profiler for laboratory and field studies. *Journal of Hydraulic Engineering, ASCE* **123**: 1089–1098.

Lemon KP, Earl AM, Vlamakis HC, Aguilar C, Kolter R. 2008. Biofilm development with an emphasis on *Bacillus subtilis*. *Current Topics in Microbiology and Immunology* **322**: 1–16.

Lenihan HS. 1999. Physical-biological coupling on oyster reefs: how habitat structure influences individual performance. *Ecological Monographs* **69**: 251–275.

Lenssen JPM, Menting FBJ, Van der Putten WH, Blom CWPM. 1999. Effects of sediment type and water level on biomass production of wetland plant species. *Aquatic Botany* **64**: 151–165.

Leonard L, Croft AL. 2006. The effect of standing biomass on flow velocity and turbulence in Spartina alterniflora canopies. *Estuarine, Coastal and Shelf Science* **69**: 325–336.

Leonard LA, Luther ME. 1995. Flow hydrodynamics in tidal marsh canopies. *Limnology and Oceanography* **40**: 1474–1484.

Leonard LA, Reed DJ. 2002. Hydrodynamics and sediment transport through tidal marsh canopies. *Journal of Coastal Research* **SI36**: 459–469.

Lesht BM. 1980. Benthic boundary-layer velocity profiles: dependence on averaging period. *Journal of Physical Oceanography* **10**: 985–991.

Leu JM, Chan HC, Jia YF, He ZG, Wang SSY. 2008. Cutting management of riparian vegetation by using hydrodynamic model simulations. *Advances in Water Resources* **31**: 1299–1308.

Levinton J. 1995. Bioturbators as ecosystem engineers: control of the sediment fabric, inter-individual interactions and material fluxes. In: Jones CG, Lawton JH (eds) *Linking species and ecosystems*. Chapman and Hall, New York, USA. pp. 29–36.

Lewis JB, Riebel PW. 1984. The effect of substrate on burrowing in freshwater mussels (Unionidae). *Canadian Journal of Zoology* **62**: 2023–2025.

Lhermitte R. 1983. Doppler sonar observation of tidal flow. *Journal of Geophysical Research* **88**: 725–742.

Lhermitte R, Lemmin U. 1994. Open-channel flow and turbulence measurement by high-resolution Doppler sonar. *Journal of Atmospheric and Oceanic Technology* **11**: 1295–1308.

Lhermitte R, Serafin R. 1984. Pulse-to-pulse coherent signal processing techniques. *Journal of Atmospheric and Oceanic Technology* **4**: 293–308.

Li CW, Xie JF. 2011. Numerical modeling of free surface flow over submerged and highly flexible vegetation. *Advances in Water Resources* **34**: 468–477.

Li CW, Yan K. 2007. Numerical investigation of wave-current-vegetation interaction. *Journal of Hydraulic Engineering* **133**: 794–803.

Li CW, Yu LH. 2010. Hybrid LES/RANS modelling of free surface flow through vegetation. *Computers and Fluids* **39**: 1722–1732.

Li CW, Zhang ML. 2010. 3D modelling of hydrodynamics and mixing in a vegetation field under waves. *Computers and Fluids* **39**: 604–614.

Li RM, Shen HW. 1973. Effect of tall vegetations on flow and sediment. *Journal of the Hydraulics Division of the American Society of Civil Engineers* **99**: 793–814.

Li X-N, Songa H-L, Li W, Lua X-W, Nishimura O. 2010. An integrated ecological floating-bed employing plant, freshwater clam and biofilm carrier for purification of eutrophic water. *Ecological Engineering* **36**: 382–390.

Li P, Liu Z, Xu R. 2001. Chemical characterisation of the released polysaccharide from the cyanobacterium Aphanothece halophytica GR02. *Journal of Applied Phycology* **13**: 71–77.

Lim JL, DeMont ME. 2009. Kinematics, hydrodynamics and force production of pleopods suggest jet-assisted walking in the American lobster (*Homarus americanus*). *The Journal of Experimental Biology* **212**: 2731–2745.

Lin B, Falconer RA. 1997. Three-dimensional layer-integrated modeling of estuarine flows with flooding and drying. *Estuarine, Coastal and Shelf Science* **44**: 737–751.

Lindstrom M, Sandberg-Kilp E. 2008. Breaking the boundary – the key to bottom recovery? The role of mysid crustaceans in oxygenizing bottom sediments. *Journal of Experimental Marine Biology and Ecology* **354**: 161–168.

Lintas C, Seed R. 1990. Spatial variation in the fauna associated with *Mytilus edulis* on a wave-exposed rocky shore. *Journal of Molluscan Studies* **60**: 165–174.

Liscum E, Stowe-Evans EL. 2000. Phototropism: a 'simple' physiological response modulated by multiple interacting photosensory-response pathways. *Photochemistry and Photobiology* **72**: 273–282.

Liu D, Diplas P, Hodges CC, Fairbanks JD. 2010. Hydrodynamics of flow through double layer rigid vegetation. *Geomorphology* **116**: 286–296.

Liu Y, Lam MC, Hang HHP. 2001. Adsorption of heavy metals by EPS of activated sludge. Water Science. *Technology* **43**(6), 59–66.

Liu Y, Tay JH. 2002. The essential role of hydrodynamic shear force in the formation of biofilm and granular sludge. *Water Research* **36**: 1653–1665.

Llewellyn DW, Shaffer GP. 1993. Marsh restoration in the presence of intense herbivory: the role of *Justicia lanceolata* (Chapm.) Small. *Wetlands* **13**: 176–184.

Lohrer AM, Thrush SF, Gibbs MM. 2004. Bioturbators enhance ecosystem function through complex biogeochemical interactions. *Nature* **431**: 1092–1095.

Lohrmann A, Cabrera R, Kraus NC. 1995. Acoustic-Doppler velocimeter (ADV) for laboratory use. In: Pugh CA. (ed) *Proceedings of Conference on Fundamentals and Advancements in Hydraulic Measurements and Experimentation, August 1994*, American Society of Civil Engineers, Buffalo, NY, USA. pp. 351–365.

López F, García MA. 1998. Open channel flow through simulated vegetation: suspended sediment transport modeling. *Water Resources Research* **34**: 2341–2352.

López F, García MH. 2001. Mean flow and turbulence structure of open-channel flow through non-emergent vegetation. *Journal of Hydraulic Engineering, ASCE* **127**(5): 392–402.

Løvås SM. 2000. *Hydro-physical conditions in kelp forests and the effect on wave damping and dune erosion-a case study on* Laminaria hyperborea. PhD Thesis, Norwegian University of Science and Technology, Norway.

Løvås SM, Tørum A. 2001. Effect of the kelp *Laminaria hyperborea* upon sand dune erosion and water particle velocities. *Coastal Engineering* **44**: 37–63.

Lövstedt CB, Larson M. 2010. Wave damping in reed: field measurements and mathematical modelling. *Journal of Hydraulic Engineering* **136**: 222–233.

Lowe RJ, Falter JL, Koseff JR, Monismith SG, Atkinson MJ. 2007. Spectral wave flow attenuation within submerged canopies: Implications for wave energy dissipation. *Journal of Geophysical Research – Oceans* **112**: C05018.

Lowe RJ, Koseff JR, Monismith SG. 2005. Oscillatory flow through submerged canopies: 1. Velocity structure. *Journal of Geophysical Research – Oceans* **110**: C10.

Lowrance RR, McIntyre S, Lance C. 1988. Erosion and deposition in a field/forest system estimated using Cesium-137 activity. *Journal of Soil and Water Conservation* **43**: 195–199.

Lu YY, Lueck RG, Huang DY. 2000. Turbulence characteristics in a tidal channel. *Journal of Physical Oceanography* **30**: 855–867.

Luce JJ, Cattaneo A, Lapointe MF. 2010a. Spatial patterns in periphyton biomass after low-magnitude flow spates: geomorphic factors affecting patchiness across gravel–cobble riffles. *Journal of the North American Benthological Society* **29**: 614–626.

Luce JJ, Steele R, Lapointe MF. 2010b. A physically based statistical model of sand abrasion effects on periphyton biomass. *Ecological Modelling* **221**: 353–361.

Luckenbach MW. 1986. Sediment stability around animal tubes: the roles of hydrodynamic processes and biotic activity. *Limnology and Oceanography* **31**: 779–787.

Luhar M, Coutu S, Infantes E, Fox S, Nepf HM. 2010. Wave-induced velocities inside a model seagrass bed. *Journal of Geophysical Research – Oceans* **115**: C12005.

Luhar M, Nepf HM. 2011. Flow-induced reconfiguration of buoyant and flexible aquatic vegetation. *Limnology and Oceangraphy* **56**: 2003–2017.

Lumborg U, Andersen TJ, Pejrup M. 2006. The effect of *Hydrobia ulvae* and microphytobenthos on cohesive sediment dynamics on an intertidal mudflat described by means of numerical modeling. *Estuarine, Coastal and Shelf Science* **68**: 208–220.

Lyautey E, Jackson CR, Cayrou J, Rols J-L, Garabétian F. 2005. Bacterial Community Succession in Natural River Biofilm Assemblages. *Microbial Ecology* **50**: 589–601.

López F, García MH. 2001. Mean flow and turbulence structure of open-channel flow through non-emergent vegetation. *Journal of Hydraulic Engineering, ASCE* **127**: 392–402.

Machata-Wenninger C, Janauer G. 1991. The measurement of current velocities in macrophyte beds. *Aquatic Botany* **39**: 221–230.

Mallik AU, Rasid H. 1993. Root-shoot characteristics of riparian plants in a flood control channel: implications for bank stabilization. *Ecological Engineering* **2**: 149–158.

Mamo M, Bubenzer, GD. 2001a. Detachment rate, soil erodibility and soil strength as influenced by living plant roots: Part II. Field study. *American Society of Agricultural Engineers* **44**: 1175–1181.

Mamo M, Bubenzer GD. 2001b. Detachment rate, soil erodibility and soil strength as influenced by living plant roots: Part I. Laboratory study. *American Society of Agricultural Engineers* **44**: 1167–1174.

Manca E. 2010. *Effects of Posidonia oceanica seagrass on nearshore waves and wave-induced flows.* PhD thesis. University of Southampton, Southampton. National Oceanography Centre, UK.

Martiny AC, Jorgensen TM, Albrechtsen HJ, Arvin E, Molin S. 2003. Long-term succession of structure and diversity of a biofilm formed in a model drinking water distribution system. *Applied Environmental Microbiology* **69**(11), 6899–6907.

Matisoff G, Fisher JB, Matis S. 1985. Effects of benthic macroinvertebrates on the exchange of solutes between sediments and freshwater. *Hydrobiologia* **122**: 19–33.

Matthaei CD, Guggelberger C, Huber H. 2003. Local disturbance history affects patchiness of benthic river algae. *Freshwater Biology* **48**: 1514–1526.

Maude SH, Williams DD. 1983. Behaviour of crayfish in water currents: Hydrodynamics of eight species with reference to their distribution patterns in southern Ontario. *Canadian Journal of Fisheries and Aquatic Sciences* **40**: 68–77.

Mayer MS, Schaffner L, Kemp WM. 1995. Nitrification potentials of benthic macrofaunal tubes and burrow walls: Effects of sediment NH sub(4) super (+) and animal behaviour. *Marine Ecology Progress Series* **121**: 157–169.

Mazda Y, Kanazawa N, Wolanski E. 1995. Tidal asymmetry in mangrove creeks. *Hydrobiologia* **295**: 51–58.

Mazda Y, Wolanski E, King BA, Sase A, Ohtsuka D, Magi M. 1997. Drag force due to vegetation in mangrove swamps. *Mangroves and Salt Marshes* 1: 193–199.

Mazik K, Curtis N, Fagan MJ, Taft S, Elliot M. 2008. Accurate quantification of the influence of benthic macro- and meio-fauna on the geometric properties of estuarine muds by micro computer tomography. *Journal of Experimental Marine Biology and Ecology* 354: 192–201.

McBride GB, Tanner CC. 2000. Modelling biofilm nitrogen transformations in constructed wetland mesocosms with fluctuating water levels. *Ecological Engineering* 14: 93–106.

McBride M, Hession WC, Rizzo DM, Thompson DM. 2007. The influence of riparian vegetation on near-bank turbulence: a flume experiment. *Earth Surface Processes and Landforms* 32: 2019–2037.

McCall PL, Tevesz MJS, Schwelgien SF. 1979. Sediment mixing by *Lamsilis radiata siliquoidea* (Mollusca) from western Lake Erie. *Journal of Great Lakes Research* 5: 105–111.

McCall PL, Tevesz MJS, Wang X, Jackson JR. 1995. Particle mixing rates of freshwater bivalves: *Anodonta grandis* (Unionidae) and *Sphaerium striatinium* (Pisidiidae). *Journal of Great Lakes Research* 21: 333–339.

McCraith BJ, Gardner LR, Wethey DS, Moore WS. 2003. The effect of fiddler crab burrowing on sediment mixing and radionuclide profiles along a topographic gradient in a southeastern salt marsh. *Journal of Marine Research* 61: 359–390.

McLelland SJ, Nicholas AP. 2000. A new method for evaluating errors in high-frequency ADV measurements. *Hydrological Processes* 14: 351–366.

McMahon RF, Bogan AE. 2001. Mollusca: Bivalvia. In: Thorp JH, Covich AP (eds) *Ecology and classification of North American freshwater invertebrates. 2nd edition.* Academic Press, Elsevier, London, UK. pp. 331–429.

Meadows PS, Shand. 1989. Experimental analysis of byssus thread production by *Mytilus edulis* and *Modiolus modoilus* in sediments. *Marine Biology* 101: 219–226.

Meadows PS, Tait J. 1989. Modification of sediment permeability and shear strength by two burrowing invertebrates. *Marine Biology* 101: 75–82.

Meadows PS, Tait J, Hussain SA. 1990. Effects of estuarine infauna on sediment stability and particle sedimentation. *Hydrobiologia* 190: 263–266.

Mendez FJ, Losada IJ. 2004. An empirical model to estimate the propagation of random breaking and nonbreaking waves over vegetation fields. *Coastal Engineering* 51: 103–118.

Mendez FJ, Losada IJ, Losada MA. 1999. Hydrodynamics induced by wind waves in a vegetation field. *Journal of Geophysical Research - Oceans* 104: 18383–18396.

Mermillod-Blondin F, Gaudet JP, Gerino M, Desrosiers G, Jose J, Crueze de Chatelliers M. 2004a. Relative influence of bioturbation and predation on organic matter processing in river sediments: a microcosm experiment. *Freshwater Biology* 49: 895–912.

Mermillod-Blondin F, Marie S, Desrosiers G, Long B, de Montety L, Michaud E, Stora G. 2003. Assessment of the spatial variability of intertidal benthic communities by axial tomodensitomtry: importance of fine-scale heterogeneity. *Journal of Experimental Marine Biology and Ecology* 287: 193–208.

Mermillod-Blondin F, Nogaro G, Datry T, Malard F, Gilbert J. 2005. Do tubificid worms influence the fate of organic matter and pollutants in stormwater sediments? *Environmental Pollution* 134: 57–69.

Mermillod-Blondin F, Rosenberg R, Francois-Carcaillet F, Norling K, Mauclaire L. 2004b. Influence of bioturbation by three benthic infaunal species on microbial communities and biogeochemical processes in marine sediment. *Aquatic Microbial Ecology* 36: 271–284.

Merz RA. 1984. Self-generated versus environmentally produced feeding currents: a comparison for the sabellid polychaete *Eudistylia vancouveri*. *Biological Bulletin* 167: 200–209.

Meyer DL, Townsend EC, Thayer GW. 1997. Stabilization and erosion control value of oyster cultch for intertidal marsh. *Restoration Ecology* 5: 93–99.

Meysman FJR, Galaktionov OS, Cook PLM, Janssen F, HUettel M, Middelburg JJ. 2007. Quantifying biologically and physically induced flow and tracer dynamics in permeable sediments. *Biogeosciences* 4: 627–646.

Michaud E, Desrosiers G, Mcrmillod-Blondin F, Sundby B, Stora G. 2005. The functional group approach to bioturbation: the effects of biodiffusers and gallery-diffusers of the *Macoma balthica* community on sediment oxygen uptake. *Journal of Experimental Marine Biology and Ecology* 326: 77–88.

Migeon S, Weber O, Faugeres JC, Saint-Paul, J. 1999. SCOPIX: A new X-ray imaging system for core analysis. *Geo-Marine Letters* 18: 251–255.

Mignot E, Barthelemy E, Hurther D. 2009. Double-averaging analysis and local flow characterization of near-bed turbulence in gravel-bed channel flows. *Journal of Fluid Mechanics* 618: 279–303, DOI:10.1017/S0022112008004643.

Milburn D, Krishnappan BG. 2003. Modelling erosion and deposition of cohesive sediments from Hay River, Northwest Territories, Canada. Paper presented at the 13th Northern Research Basins Workshop, 19th–24th August 2001, Saariselkä, Finland and Murmansk, Russia.

Miller KS, Rochwarger MM. 1972. A covariance approach to spectral moment estimation. *IEEE Transactions on Information Theory* IT-18: 588–596.

Mitsch WJ, Cronk JK, Wu X, Nairn RW, Hey DL. 1995. Phosphorus retention in constructed freshwater riparian marshes. *Ecological Applications* 5: 830–845.

Moin P, Mahesh K. 1998. Direct Numerical Simulation: A tool in turbulence research. *Annual Review of Fluid Mechanics* 30: 539–78.

Molin S, Tolker-Nielsen, T. 2003. Gene transfer occurs with enhanced efficiency in biofilms and induces enhanced stabilisation of the biofilm structure. Current Opinion in Biotechnology 14(3): 255–261.

Möller I. 2006. Quantifying saltmarsh vegetation and its effect on wave height dissipation: Results from a UK East coast saltmarsh. *Estuarine, Coastal and Shelf Science* 69: 337–351.

Möller I, Mantilla-Contreras J, Spencer T, Hayes A. 2011. Micro-tidal coastal reed beds: Hydro-morphological insights and observations on wave transformation from the southern Baltic Sea. *Estuarine, Coastal and Shelf Science* 92: 424–436.

Möller I, Spencer T, French JR, Leggett DJ, Dixon M. 1999. Wave transformation over salt marshes: a field and numerical modelling study from North Norfolk, England. *Estuarine, Coastal and Shelf Science* 49: 411–426.

Monismith SG, Koseff JR, Thompson JK, O'Riordan CA, Nepf HM. 1990. A study of model bivalve siphonal currents. *Limnology and Oceanography* 35: 680–696.

Morin J, Leclerc M, Secretan Y, Boudreau P. 2000. Integrated two dimensional macrophytes-hydrodynamic modelling. *Journal of Hydraulic Research* 38: 163–172.

Mork M. 1996. The effect of kelp in wave damping. *Sarsia* 80: 323–327.

Mork M. 1996. *Wave attenuation due to bottom vegetation, waves and nonlinear processes in hydrodynamics*. Kluwer Academic Publishing, Oslo, Norway, pp. 371–382.

Morris EP, Peralta G, Brun FG, van Duren LA, Bouma TJ, Perez-Llorens JL. 2008. Interaction between hydrodynamics and seagrass canopy structure: Spatially explicit effects on ammonium uptake rates. *Limnology and Oceanography* 53: 1531–1539.

Morton B. 1974. Some aspects of the biology, population dynamics, and functional morphology of *Musculista senhausia* Benson (Bivalvia, Mytilidae). *Pacific Science* 28: 19–33.

Moulin FY, Guizien K, Thouzeau G, Chapalain G, Mülleners K, Bourg C. 2007. Impact of an invasive species, *Crepidula fornicata*, on the hydrodynamics and transport properties of the benthic boundary layer. *Aquatic Living Resources* 20: 15–31.

Müller UK, Stamhuis EJ, Videler JJ. 2000. Hydrodynamics of unsteady fish swimming and the effects of body size: comparing the flow fields of fish larvae and adults. *Journal of Experimental Biology* 203: 193–206.

Murphy RC. 1985. Factors affecting the distribution of the introduced bivalve, Mercenaria mercenaria, in a California lagoon–the importance of bioturbation. *Journal of Marine Research* **43**: 673–692.

Murray JMH, Meadows A, Meadows PS. 2002. Biogeomorphological implications of microscale interactions between sediment geotechnics and marine benthos: a review. *Geomorphology* **47**: 15–30.

Muschenheim DK. 1987. The dynamics of near-bed seston flux and suspension-feeding benthos. *Journal of Marine Research* **45**: 473–496.

Myers AC. 1972. Tube-worm-sediment relationships of *Diopatra cuprea* (Polychaeta: Onuphidae). *Marine Biology* **17**: 350–356.

Myrhaug D, Holmedal LE, Ong MC. 2009. Nonlinear random wave-induced drag force on a vegetation field. *Coastal Engineering* **56**: 371–376.

Nagayama K, Tanaka K. 2006. 2D-PIV analysis of loach motion and flow field. *Journal of Visualization* **9**: 393–401.

Nagayama K, Tanaka T, Tanaka K, Hayami H, Aramaki S. 2008. Visualization of flow and vortex structure around a swimming loach by dynamic stereoscopic PIV. *Experiments in Fluids* **44**: 843–850.

Naot D, Nezu I, Nakagawa H. 1996. Hydrodynamic behaviour of partly vegetated open channels. *Journal of Hydraulic Engineering* **122**: 625–633.

Nates SF, Felder DL. 1998. Impacts of burrowing ghost shrimp, genus *Lepidophthalmus* (Crustacea: Decapoda: Thalassinidea), on penaeid shrimp culture. *Journal of the World Aquaculture Society* **29**: 188–210.

Neary VS. 2003. Numerical solution of fully developed flow with vegetative resistance. *Journal of Engineering Mechanics* **129**: 558–563.

Nepf HM. 1999. Drag, turbulence, and diffusion in flow through emergent vegetation. *Water Resources Research* **35**: 479–489.

Nepf HM. 2012. Flow and transport in and around aquatic vegetation. *Annual Review of Fluid Mechanics* **44**: 123–142.

Nepf HM, Sullivan JA, Zavistoski RA. 1997. A model for diffusion within emergent vegetation. *Limnology and Oceanography* **42**: 1735–1745.

Nepf HM, Vivoni ER. 2000. Flow structure in depth-limited, vegetated flow. *Journal of Geophysical Research* **105**: 28547–28557.

Neumeier U. 2005. Quantification of vertical density variations of salt-marsh vegetation. *Estuarine, Coastal and Shelf Science* **63**: 489–496.

Neumeier U. 2007. Velocity and turbulence variations at the edge of saltmarshes. *Continental Shelf Research* **27**: 1046–1059.

Neumeier U, Amos CL. 2006. Turbulence reduction by the canopy of coastal Spartina salt-marshes. *Journal of Coastal Research* **39**: 433–439.

Neumeier U, Ciavola P. 2004. Flow resistance and associated sedimentary processes in a *Spartina maritima* salt-marsh. *Journal of Coastal Research* **20**: 435–447.

Neumeier U, Friend PL, Gangelhof U, Lunding J, Lundkvist M, Bergamasco A, Amos CL, Flindt M. 2007. The influence of fish feed pellets on the stability of seabed sediment: A laboratory flume investigation. *Estuarine, Coastal and Shelf Science* **75**: 347–357.

Newell RIE, Koch EW. 2004. Modeling seagrass density and distribution in response to changes in turbidity stemming from bivalve filtration and seagrass sediment stabilization. *Estuaries* **27**: 793–806.

Newton TD, Gattie DK, Lewis DL. 1990. Initial test of the benchmark chemical approach for predicting microbial transformation rates in aquatic environments. *Applied and Environmental Microbiology* **56**: 288–291.

Nezu I, Nakagawa H. 1993. *Turbulence in Open-channel Flows*. IAHR Monograph Series. A.A. Balkema, Rotterdam, The Netherlands.

Nezu I, Onitsuka K. 2001. Turbulent structures in partly vegetated open-channel flows with LDA and PIV measurements. *Journal of Hydraulic Research* 39: 629–642.

Nezu I, Rodi W. 1986. Open-channel flow measurements with a Laser Doppler Anemometer. *Journal of Hydraulic Engineering, ASCE* 112: 335–355.

Nezu I, Sanjou M. 2008. Turbulence structure and coherent motion in vegetated open-channel flows. *Journal of Hydro-environment Research* 2: 62–90.

Nickell LA, Atkinson RJA. 1995. Functional morphology of burrows and trophic modes of three thalassinidean shrimp species, and a new approach to the classification of thalassinidean burrow morphology. *Marine Ecology Progress Series* 128: 181–197.

Nickell LA, Atkinson RJA, Hughes DJ, Ansell AD, Smith CJ. 1995. Burrow morphology of the echiuran worm *Maxmuelleria lankesteri* (Echiura: Bonelliidae), and a brief review of burrow structure and related ecology of the Echiura. *Journal of Natural History* 29: 871–885.

Nield DA, Bejan A. 1999. Convection in Porous Media, 2nd ed., Springer-Verlag, New York.

Nielsen TS, Funk WH, Gibbons HL, Duffner RM. 1984. A comparison of periphyton growth on artificial and natural substrates in the Upper Spokane River. *Northwest Science* 58: 243–248.

Nikora V, Green MO, Thrush SF, Hume TM, Goring D. 2002a. Structure of the internal boundary layer over a patch of pinnid bivalves (*Atrina zelandica*) in an estuary. *Journal of Marine Research* 60: 121–150.

Nikora VI, Goring DG, Biggs BJF. 1997. On stream periphyton-turbulence interactions. *New Zealand Journal of Marine and Freshwater Research* 31: 435–448.

Nikora VI, Goring DG, Biggs BJF. 2002b. Some observations of the effects of micro-organisms growing on the bed of an open channel on the turbulence properties. *Journal of Fluid Mechanics* 450: 317–341.

Nikora VI, Goring DG, McEwan I, Griffiths G. 2001. Spatially-averaged open-channel flow over a rough bed. *Journal of Hydraulic Engineering* 127: 123–133.

Nikora VI, McEwan I, McLean S, Coleman S, Pokrajak D, Walters R. 2007a. Double-averaging concept for rough-bed open-channel and overland flows: Theoretical background. *Journal of Hydraulic Engineering* 133: 873–883.

Nikora VI, McLean S, Coleman S, Pokrajak D, McEwan I, Campbell L, Aberle J, Clunie D, Koll K. 2007b. Double-averaging concept for rough-bed open-channel and overland flows: Applications. *Journal of Hydraulic Engineering* 133: 884–895.

Nikuradse J. 1933. Strömungsgesetze in Rauhen Rohren. *VDI Forschungsheft* 4.

Noe GB, Childers DL, Edwards AL, Gaiser E, Jayachandran K, Lee D, Meeder J, Richards J, Noernberg MA, Fourner J, Dubois S, Populus J. 2010. Using airborne laser altimetry to estimate Sabellaria alveolata (Polychaeta: Sabellariidae) reefs volume in tidal flat environments. *Estuarine, Coastal and Shelf Science* 90: 93–102.

Nogaro G, Mermillod-Blondin F, Francois-Carcaillet F, Gaudet JP, LaFont M, Gibert J. 2006. Invertebrate bioturbation can reduce the clogging of sediment: an experimental study using infiltration sediment columns. *Freshwater Biology* 51: 1458–1473.

Norkko A, Hewitt JE, Thrush SF, Funnell GA. 2006. Conditional outcomes of facilitation by a habitat-modifying subtidal bivalve. *Ecology* 87: 226–234.

Nortek. 1997. *ADV Operation Manual*. Nortek AS, Bruksveien 17, 1390 Vollen, Norway, 33.

Nowell ARM, Church M. 1979. Turbulent flow in a depth-limited boundary layer. *Journal of Geophysical Research* 84: 4816–4824

Nowell ARM, Jumars PA, Eckman JE. 1981. Effects of biological activity on the entrainment of marine sediments. *Marine Geology* 42: 133–153.

Nowell ARM, Jumars PA. 1984. Flow environments of aquatic benthos. *Annual Review of Ecology and Systematics* **15**: 303–328.

Nowell ARM, Jumars PA. 1987. Flumes: Experimental and theorectiical experimental considerations for simulation of benthic environments. *Oceanography and Marine Biology: An Annual Review* **25**: 91–112.

Nyström P. 2005. Non-lethal predator effects on the performance of a native and an exotic crayfish species. *Freshwater Biology* **50**: 1938–1949.

O'Donnell MJ. 2008. Reduction of wave forces within bare patches in mussel beds. *Marine Ecology Progress Series* **362**: 157–167.

O'Reilly R, Kennedy R, Patterson A. 2006. Destruction of conspecific bioturbation structures by *Amphiura filiformis* (Ophiuroida): evidence from luminophore tracers and *in situ* time-lapse sediment-profile imagery. *Marine Ecology Progress Series* **315**: 99–11.

O'Riordan CA, Monismith SG, Koseff JR. 1993. A study of concentration boundary-layer formation over a bed of model bivalves. *Limnology and Oceanography* **38**: 1712–1729.

O'Riordan CA, Monismith SG, Koseff JR. 1995. The effect of bivalve excurent jet dynamics on mass transfer in a benthic boundary layer. *Limnology and Oceanography* **40**: 330–344.

O'Toole G, Kaplan HB, Kolter R. 2000. Biofilm formation as microbial development. *Annual Review of Microbiology* **54**: 49–79.

O'Donnel MJ. 2008. Reduction of wave forces within bare patches in mussel beds. *Marine Ecology Progress Series* **362**: 157–167.

Ofalsson E. 2003 Do macrofauna structure meiofauna assemblages in marine soft-bottoms? A review of experimental studies. *Vie Milieu* **53**: 249–265.

Olesen B, Madsen TV. 2000. Growth and physiological acclimation to temperature and inorganic carbon availability by two submerged aquatic macrophyte species, *Callitriche cophocarpa* and *Elodea canadensis. Functional Ecology* **14**: 252–260

Ortega-Morales BO, Santiago-Garcia JL, Chan-Bacab MJ, Moppert X, Miranda-Tello E, Fardeau ML, Carrero JC, Bartolo-Perez P, Valadez-Gonzalez A, Guezennec J. 2006. Characterization of extracellular polymers synthesized by tropical intertidal biofilm bacteria. *Journal of Applied Microbiology* **102**: 254–264.

Orvain F. 2005. A model of sediment transport under the influence of surface bioturbation: generalisation to the facultative suspension-feeder *Scrobicularia plana. Marine Ecology Progress Series* **286**: 43–56.

Orvain F, Le Hir P, Sauriau P-G. 2003b. A model of fluff layer erosion and subsequent bed erosion in the presence of the bioturbator, *Hydrobia ulvae. Journal of Marine Research* **61**: 823–851.

Orvain F, Sauriau P-G, Bacher C, Prineau M. 2006. The influence of sediment cohesiveness on bioturbation effects due to *Hydrobia ulvae* on the initial erosion of intertidal sediments: a study combining flume and model approaches. *Journal of Sea Research* **55**: 54–73.

Orvain F, Sauriau P-G, Sygut A, Joassard L, Le Hir P. 2004. Interacting effects of *Hydrobia ulvae* bioturbation and microphytobenthos on the erodibiility of mudflat sediments. *Marine Ecology Progress Series* **278**: 205–223.

Osmundson DB, Ryel RJ, Lamarra VL, Pitlick J. 2002. Flow-sediment-biota relations: implications for river regulation effects on native fish abundance. *Ecological Applications* **12**: 1719–1739.

Paine RT, Levin SA. 1981. Intertidal landscapes: disturbance and the dynamics of pattern. *Ecological Monograph* **51**: 145–178.

Paola C. 2001. Modelling stream braiding over a range of scales. In: Mosley MP. (ed) *Gravel-Bed Rivers V.* New Zealand Hydrological Society, Wellington, New Zealand. pp. 11–46.

Pardo R, Herguedas M, Barrado E, VegaM. 2003. Biosorption of cadmium, copper, lead and zinc by inactive biomass of *Pseudomonas putida. Anal Bioanal Chem* **376**: 26–32.

Pasche E, Rouvé G. 1985. Overbank flow with vegetatively roughened flood plains. *Journal of Hydraulic Engineering, ASCE* **111**: 1262–1278.

Passarelli C, Hubas C, Segui AN, Grange J, Meziane T. 2012. Surface adhesion of microphytobenthic biofilms is enhanced under Hediste diversicolor (O.F. Müller) trophic pressure. Journal of Experimental Marine Biology and Ecology **438**: 52–60.

Paterson DM, Tolhurst TJ, Kelly JA, Honeywill C, de Dekere EMGT, Huet V, Shayler SA, Black KS, de Brouwer J, Davidson I. 2000. Variations in sedimentproperties, Skeffling mudflat, Humber Estuary, UK. *Continental Shelf Research* **20**: 1373–1396.

Patton EG, Shaw RH, Judd MJ, Raupach MR. 1998. Large-eddy simulation of windbreak flow. *Boundary-Layer Meteorology* **87**: 257–306.

Paul M. 2011. *The role of* Zostera noltii *in wave attenuation.* PhD thesis. University of Southampton, Southampton. National Oceanography Centre, UK.

Paul M, Amos CL. 2011. Spatial and seasonal variation in wave attenuation over *Zostera noltii. Journal of Geophysical Research* **116**: C08019.

Paul M, Bouma TJ, Amos CL. 2012. Wave attenuation by submerged vegetation: combining the effect of organism traits and tidal current. *Marine Ecology Progress Series* **444**: 31–41. DOI:10.3354/meps09489

Peakall J, Amos KJ, Keevil GM, Bradbury PW, Gupta S. 2007. Flow processes and sedimentation in submarine channel bends. *Marine and Petroleum Geology* **24**: 470–486.

Peakall J, Ashworth P, Best JL. 1996. Physical modelling in fluvial geomorphology: principles, applications and unresolved issues. In: Rhoads BL, Thorn CE (eds) *The Scientific Nature of Geomorphology.* John Wiley and Sons, Chichester, UK. pp. 221–253.

Peine F, Bobertz B, Graf G. 2011. Influence of the blue mussel *Mytilus edulis* (Linnaeus) on the bottom roughness length (z_0) in the south-western Baltic Sea. *Baltica* **18**: 13–22.

Peine F, Friedrichs M, Graf G. 2009. Potential influence of tubicolous worms on the bottom roughness length z_0 in the south-western Baltic Sea. *Journal of Experimental Marine Biology and Ecology* **374**: 1–11.

Pei-shi QI, Wen-bin W, Zheng, QI 2008. Effect of shear stress on biofilm morphological characteristics and the secretion of extracellular polymeric substances. In proceeding of Bioinformatics and Biomedical Engineering pp. 3438–3441.

Pemberton GS, Risk MJ, Buckley DE. 1976. Supershrimp: Deep bioturbation in the Strait of Canso, Nova Scotia. *Science* **192**: 790–791.

Penning WE, Raghuraj R, Mynett A. 2009. The effects of macrophyte morphology and patch density on wave attenuation. 7th International Symposium on Ecohydraulics Conference, 12th–16th January 2009. Concepción, Chile.

Perez KT, Davey EW, Moore RH, Burn PR, Rosol MS, Cardin JA, Johnson RL, Kopans DN. 1999. Application of computer-aided tomography (CT) to the study of estuarine benthic communities. *Ecological Application* **9**: 1050–1058.

Petersen JK, Riisgård HU. 1992. Filtration capacity of the ascidian *Ciona intestinalis* and its grazing impact in a shallow fjord. *Marine Ecology Progress Series* **88**: 9–17.

Peterson CG. 1996. Response of benthic algal communities to natural physical disturbance. In: Stevenson RJ, Bothwell ML, and Lowe RL. (ed) *Algal Ecology: Freshwater Benthic Ecosystems.* Academic, San Diego, USA. pp. 375–403.

Peterson CH, Luettich RA, Micheli F, Skilleter GA. 2004. Attenuation of water flow inside seagrass canopies of differing structure. *Marine Ecology Progress Series* **268**: 81–92.

Pethick J, Burd F. 1993. *Coastal Defence and the Environment: a Guide to Good Practice.* Ministry of Agriculture, Fisheries and Food, UK.

Petryk S, Bosmajian G. 1975. Analysis of flow through vegetation. *Journal of the Hydraulics Division of the American Society of Civil Engineers* **101**: 871–884.

Pham, N., Penning, E., Mynett, A., Raghuraj, R., 2011. Effects of submerged tropical macrophytes on flow resistance and velocity profiles in open channels. *International Journal of River Basin Management* 9 (3-4), 195–203. doi: 10.1080/15715124.2011.648775.

Pietri L, Amielh M, Anselmet F. 2006. Effect of the vegetation density on the turbulence properties in a canopy flow. 13th International Symposium on Applying Laser Techniques to Fluid
Mechanics, Lisbon, Portugal.

Pietri L, Petroff A, Amielh M, Anselmet F. 2009. Turbulent flows interacting with varying density canopies. *Mécanique and Industries* 10: 181–185.

Pilditch CA, Grant J. 1999. Effect of variations in flow velocity and phytoplankton concentration on sea scallop (*Placopecten magellanicus*) grazing rates. *Journal of Experimental
Marine Biology and Ecology* 240: 111–136.

Pitlo RH, Dawson FH. 1990. Flow resistance of aquatic weeds. In: Pieterse AH, Murphy
KJ. (eds) *Aquatic Weeds: The Ecology and Management of Nuisance Aquatic Vegetation.*
Oxford University Press, Oxford, pp. 74–84.

Plew DR, Cooper GG, Callaghan FM. 2008. Turbulence induced forces in a freshwater macrophyte canopy. *Water Resources Research* 44.

Plew DR, Enright MP, Nokes RI, Dumas JK. 2009. Effect of mussel bio-pumping on the drag
and flow around a mussel crop rope. *Aquacultural Engineering* 40: 55–61.

Poggi D, Porporato A, Ridolfi L, Albertson D-J, Katul G-G. 2004. The effect of vegetation density on canopy sub-layer turbulence. *Boundary-Layer Meteorology* 111: 565–587.

Pollen N, Simon A. 2005. Estimating the mechanical effects of riparian vegetation on streambank stability using a fiber bundle model. *Water Resources Research* 41: W07025,
DOI:10.1029/2004 WR003801.

Pollen N, Simon A. 2006. Geotechnical implications for the use of alfalfa in experimental studies of alluvial-channel morphology and planform. *Eos Transactions, American Geophysical
Union (AGU)* 87: Fall Meeting Supplement, Abstract H21G–1455.

Pollen-Bankhead N, Thomas RE, Gurnell AM, Liffen T, Simon A, O'Hare MT. 2011. Quantifying the potential for flow to remove the emergent aquatic macrophyte *Sparganium erectum* from the margins of low-energy rivers. *Ecological Engineering* 37: 1779–1788.

Pope SB. 2000. *Turbulent Flows.* Cambridge University Press, Cambridge, UK.

Porter ET, Sanford LP, Suttles SE. 2000. Gypsum dissolution is not a universal integrator of
'water motion'. *Limnology and Oceanography* 45: 145–158.

Posey MH, Pregnall AM, Graham RA. 1984. A brief description of a subtidal Sabellariid (Polychaeta) reef on the southern Oregon Coast. *Pacific Science* 38: 28–33.

Powell EN, Staff GM, Davies DJ, Callender WR. 1989. Macrobenthic death assemblages in
modern marine environments: formation, interpretation, and application. *Critical Review
of Aquatic Science* 1: 555–589.

Powell GVN, Schaffner FC. 1991. Water trapping by seagrasses occupying bank habitats in
Florida Bay. *Estuarine, Coastal and Shelf Science* 32: 43–60.

Pratolongo PD, Perillo GME, Piccolo MC. 2010. Combined effects of waves and plants on a
mud deposition event at a mudflat-saltmarsh edge in the Bahía Blanca estuary. *Estuarine,
Coastal and Shelf Science* 87: 207–212.

Precht E, Janssen F, Huettel M. 2006. Near-bottom performance of the acoustic Doppler velocimeter ADV – A comparative study. *Aquatic Ecology* 40: 481–492.

Preston J H. 1954. The determination of turbulent skin friction by means of Pitot tubes. *Journal
of the Royal Aeronautical Society* 58: 109–121.

Price HA. 1981. Byssus thread strength in the mussel *Mytilus edulis*. *Journal of the Zoology*
194: 245–255.

Price RE, Schiebe FR. 1978. Measurements of velocity from excurrent siphons of freshwater
clams. *The Nautilus* 92: 67–69.

Price WA, Tomlinson KW, Hunt JN. 1968. The effect of artificial seaweed in promoting the
buildup of beaches. In: 11th Coastal Engineering Conference, pp. 570–578.

Pringle CM. 1985. Effects of chironomid (Insecta: Diptera) tube-building activities on stream
diatom communities. *Journal of Phycology* 21: 185–194.

Prinos P, Stratigaki V, Manca E, Losada IJ, Lara JL, Sclavo M, Caceres I, Sanchez-Arcilla A. 2010. Wave propagation over Posidonia oceanica: Large scale experiments. In: Grüne J, Breteler MK (eds) *Hydralab III Joint Transnational Access User Meeting*. Hannover. pp. 57–60.

Przeslawski R, Dundas K, Radke L, Anderson TJ. 2012. Deep-sea Lebensspuren of the Australian continental margins, Deep-Sea Research I. **65**: 26–35

Przeslawski R, Zhu Q, Aller R. 2009. Effects of abiotic stressors on infaunal burrowing and associated sediment characteristics. *Marine Ecology Progress Series* **392**: 33–42.

Puijalon S, Bornette G, Sagnes P. 2005. Adaptations to increasing hydraulic stress: morphology, hydrodynamics and fitness of two higher aquatic plant species. *Journal of Experimental Botany* **56**: 777–786.

Puijalon S, Lena J-P, Bornette G. 2007. Interactive effects of nutrient and mechanical stresses on plant morphology. *Annals of Botany* **100**: 1297–1305.

Puijalon S, Léna JP, Riviére N, Champagne JY, Rostan JC, Bornette G. 2008. Phenotypic plasticity in response to mechanical stress: hydrodynamic performance and fitness of 4 aquatic plant species. *New Phytologist* **177**: 907–917

Pullen J, LaBarabera M. 1991. Modes of feeding in aggregations of barnacles and shape of aggregations. *Biological Bulletin* **181**: 442–452.

Quintana CO, Tang M, Kristensen E. 2007. Simultaneous study of particle reworking, irrigation transport and reaction rates in sediment bioturbated by the polychaetes *Heteromastus* and *Marenzelleria*. *Journal of Experimental Marine Biology and Ecology* **352**: 392–406.

Raffel M, Willert C, Kompenhans J. 1998. *Particle Image Velocimetry: A Practical Guide*. Springer Verlag, Berlin, Germany.

Ragnarsson SA, RaVaelli D. 1999. Effects of the mussel *Mytilus edulis* L. on the invertebrate fauna of sediments. *Journal of Experimental Marine Biology and Ecology* **241**: 31–43.

Raupach M, Antonia R, Rajagopalan S. 1991. Rough-wall turbulent boundary layers. *Applied Mechanics Reviews* **44**: 1–25.

Raupach MR, Shaw RH. 1982. Averaging procedures for flow within vegetation canopies. *Boundary-Layer Meteorology* **22**: 79–90.

Rees CP. 1976. Sand grain size distribution in tubes of *Sabellara vulgaris* Verrill. *Chesapeake Science* **17**: 59–61.

Reynolds O. 1895. On the dynamical theory of incompressible viscous fluids and the determination of the criterion. *Philosophical Transactions of the Royal Society of London A* **186**: 123–164.

Rhoads DC. 1974. Organism-sediment relations on the muddy sea floor. *Oceanography and Marine Biology: an Annual Review* **12**: 263–300.

Rhoads DC, Young DK. 1970a. Animal-Sediment relations in Cape Cod Bay, Massachusetts I. Reworking by Molpadia oolitica (Holothuroidea). *Marine Biology* **11**: 255–261.

Rhoads DC, Young DK. 1970b. The influence of deposit feeding organisms on sediment stability and community trophic structure. *Journal of Marine Research* **28**: 150–178.

Rice DL. 1986. Early diagenesis in bioadvective sediments: relationships between the diagenesis of beryllium-7, sediment reworking rates, and the abundance of conveyer-belt deposit feeders. *Journal of marine Research* **44**: 149–184.

Rice SP, Buffin-Bélanger T, Lancaster J, Reid I. 2008. Movements of a macroinvertebrate (Potamophylax latipennis) across a gravel-bed substrate: effects of local hydraulics and micro-topography under increasing discharge. In: Habersack H, Piegay H, Rinaldi M. (eds) Gravel-bed Rivers VI: From Process Understanding to River Restoration. Elsevier, Amsterdam, The Netherlands: 637–660.

Rice SP, Johnson MF, Reid I. 2012. Animals and the geomorphology of gravel-bed rivers. In: Church M, Biron P, Roy A. (eds) Gravel-bed Rivers: Processes, Tools, Environments. John Wiley and Sons, New York, USA: 225–240.

Rice SP, Lancaster J, Kemp P. 2010a. Experimentation at the interface of fluvial geomorphology, stream ecology and hydraulic engineering and the development of an effective, interdisciplinary river science. *Earth Surface Processes and Landforms* **35**: 64–77.

Richardson CA, Ibarrola I, Ingham RJ. 1993. Emergence pattern and spatial distribution of the common cockle *Cerastoderma edule*. *Marine Ecology Progress Series* **99**: 71–81.

Richardson EV, Davis SR. 2001. Evaluating Scour at Bridges, Fourth Edition. *Hydraulic Engineering Circular* **18**, Publication Number FHWA NHI 01–001. Federal Highway Administration, U.S. Department of Transportation, Washington DC, USA.

Rickard AH, Gilbert P, High NJ, Kolenbrander PE, Handley PS. 2003. Bacterial coaggregation: an integral process in the development of multi-species biofilms. *Trends in Microbiology* **11**: 94–100.

Rickard AH, McBain AJ, Stead AT, Gilbert P. 2004. Shear rate moderates community diversity in freshwater biofilms. *Applied Environmental Microbiology* **70**(12), 7426–7435.

Riddle MJ. 1988. Cyclone and bioturbation effects on sediments from coral reef lagoons. *Estuarine and Coastal Marine Science* **27**: 687–695.

Righetti M. 2008. Flow analysis in a channel with flexible vegetation using double-averaging method. *Acta Geophysica* **56**: 801–823.

Righetti M, Armanini A. 2002. Flow resistance in open channel flows with sparsely distributed bushes. *Journal of Hydrology* **269**: 55–64.

Righetti M, Lucarelli C. 2010. Resuspension phenomena of benthic sediments: the role of cohesion and biological adhesion. *River Research and Applications* **26**: 404–413.

Riisgård HU. 1991. Suspension feeding in the polychaete *Nereis diversicolor*. *Marine Ecology Progress Series* **70**: 29–37.

Riisgård HU, Banta GT. 1998. Irrigation and deposit feeding by the lugworm *Arenicola marina*, characteristics and secondary effects on the environment. A review of current knowledge. *Vie Milieu* **48**: 243–257.

Riisgård HU, Kittner C, Seerup DF. 2003. Regulations of opening state and filtration rate in filter-feeding bivalves (*Cardium edule, Mytilus edulis, Mya arenaria*) in response to low algal concentrations. *Journal of Experimental Marine Biology and Ecology* **284**: 105–127.

Riisgård HU, Larsen PS. 2000. Comparative ecophysiology of active zoobenthic filter feeding, essence of current knowledge. *Journal of Sea Research* **44**: 169–193.

Rochex A, Godon JJ, Bernet N, Escudie R. 2008. Role of shear stress on composition, diversity and dynamics of biofilm bacterial communities. *Water Resources* **42**: 4915–4922.

Roditi HA, Caraco NF, Cole JJ, Strayer DL. 1996. Filtration of Hudson River water by the Zebra mussel (*Dreissena polymorpha*). *Estuaries* **19**: 824–832.

Roeselers G, van Loosdrecht M, Muyzer G. 2007. Heterotrophic pioneers facilitate phototrophic biofilm development. *Microbial Ecology* **54**: 578–585.

Rogerson M, Pedley HM, Middleton R. 2010. Microbial Influence on Macroenvironment Chemical Conditions in Alkaline (Tufa) Streams; Perspectives from In Vitro Experiments. In Speleothems and Tufas: Unravelling Physical and Biological controls, H.M. Pedley and M. Rogerson, Editors, Geological Society of London: London. pp. 65–81

Rolland T, Lemmin U. 1997. A two-component Acoustic Velocity Profiler for use in turbulent open-channel flow. *Journal of Hydraulic Research* **35**: 545–561.

Romaní AM, Sabater S. 1999. Effect of primary producers on the heterotrophic metabolism of a stream biofilm. *Freshwater Biology* **41**: 729–736

Rominger J, Lightbody A, Nepf H. 2010. Effects of added vegetation on sand bar stability and stream hydrodynamics. *J. Hydraulic Eng.* **136**(12), 994–1002.

Rosenberg R, Davey E, Gunnarsson J, Norling K, Frank M. 2007. Application of computer-aided tomography to visualize and quantify biogenic structures in marine sediments. *Marine Ecology Progress Series* **331**: 23–34.

Rosenberg R, Grémare A, Duchêne JC, Davey E, Franck M. 2008. 3D visualization and quantification of marine benthic biogenic structures and particle transport utilizing computer-aided tomography. *Marine Ecology Progress Series* 363: 171–182.

Rosenberg R, Ringdahl K. 2005. Quantification of biogenic 3-D structures in marine sediments. *Journal of Experimental Marine Biology and Ecology* 326: 67–76.

Roskosch A, Morad MR, Khalili A, Lewandowski J. 2010. Bioirrigation by *Chironomus plumosus*: advective flow investigated by particle image velocimetry. *Journal of the North American Benthological Society* 29: 789–802.

Rowden AA, Jago CF, Jones SE. 1998b. Influence of benthic macrofauna on the geotechnical and geophysical properties of surficial sediment, North Sea. *Continental Shelf Research* 18: 1347–1363.

Rowden AA, Jones MB. 1993. Critical evaluation of sediment turnover estimates for Callianassidae (Decapoda: Thalassinidea). *Journal of Experimental Marine Biology and Ecology* 173: 265–272.

Rowden AA, Jones MB. 1994. A contribution to the biology of the burrowing mud shrimp, *Callianassa subterranea* (Decapoda: Thalassinidea). *Journal of the Marine Biological Association of the UK* 74: 623–635.

Rowden AA, Jones MB, Morris AW. 1998a. The role of *Callianassa subterranea* (Montagu) (Thalassinidea) in sediment resuspension in the North Sea. *Continental shelf research* 18: 1365–1380.

Rowinski PM, Kubrak J. 2003. A mixing-length model for predicting vertical velocity distribution in flows through emergent vegetation. *Hydrological Sciences* 47: 893–904.

Rudnicki M, Mitchell SJ, Novak MD. 2004. Wind tunnel measurements of crown streamlining and drag relationships for three conifer species. *Canadian Journal of Forest Research* 34: 666–676.

Runck C. 2007. Macroinvertebrate production and food web energetic in an industrially contaminated stream. *Ecological Applications* 17: 740–753.

Rundle SD, Spicer JI, Coleman RA, Vosper J, Soane J. 2004. Environmental calcium modifies induced defences in snails. *The Royal Society* 271: 67–70.

Sakakibara J, Nakagawa M, Yoshida M. 2004. Stereo-PIV study of flow around a maneuvering fish. *Experiments in Fluids* 36: 282–293.

Salant NL. 2011. 'Sticky business': The influence of streambed periphyton on particle deposition and infiltration. *Geomorphology* 126: 350–363.

Salleh, S. and McMinn, A. 2011. The effects of temperature on the photosynthetic parameters and recovery of two temperate benthic microalgae, amphora cf. coffeaeformis and cocconeis cf. sublittoralis (bacillariophyceae). *Journal of Phycology* 47: 1413–1424.

Sand-Jensen K. 2003. Drag and reconfiguration of freshwater macrophytes. *Freshwater Biology* 48: 271–283.

Sand-Jensen K. 2008. Drag forces on common plant species in temperate streams: consequences of morphology, velocity and biomass. *Hydrobiologia* 610: 307–319.

Sand-Jensen K, Mebus JR. 1996. Fine-scale patterns of water velocity within macrophyte patches in streams. *Oikos* 76: 169–180.

Sand-Jensen K, Pedersen O. 1999. Velocity gradients and turbulence around macrophyte stands in streams. *Freshwater Biology* 42: 315–328.

Santamaria L, van Vierssen W. 1997. Photosynthetic temperature responses of fresh- and brackish-water macrophytes: a review. *Aquatic Botany* 58: 135–150.

Saunders PV. 2012. Can tufa Mg/Ca ratios be used as a palaeoclimate proxy? PhD thesis, University of Hull, UK.

Saurel C, Gascoigne JC, Palmer MR, Kaiser MJ. 2007. In situ mussel feeding behaviour in relation to multiple environmental factors: regulation through food concentration and tidal conditions. *Limnology and Oceanography* 52: 1919–1929.

Schanz A, Asmus H. 2003. Impact of hydrodynamics on development and morphology of inter-tidal seagrasses in the Wadden Sea. *Marine Ecology Progress Series* **261**: 123–134.

Schindler RJ, Lane SN, Keylock CJ, Naden PS. 2004. Characterisation of stem wake effects using PIV imagery. In: Jirka GH, Uijttewaal WSJ. (eds) *Shallow Flows*, Balkema, Rotter-dam, Netherlands. pp. 275–285.

Schoneboom T, Aberle J, Dittrich A. 2010. Hydraulic resistance of vegetated flows: Contribu-tion of bed shear stress and vegetative drag to total hydraulic resistance. In: Dittrich A, Koll K, Aberle J, Geisenhainer P. (eds) *River Flow 2010*. Bundesanstalt für Wasserbau, Karlsruhe, Germany. pp. 269–276.

Schoneboom T, Aberle J, Wilson CAME, Dittrich A. 2008. Drag force measurements of vegeta-tion elements. In: *Proceedings of ICHE-2008, Nagoya, Japan, September 8-12 2008*. ICHE-2008 Conference Secretariat, Nagoya, Japan.

Schulz M, Kozerski H, Pluntke T, Rinke K. 2003. The influence of macrophytes on sedimen-tation and nutrient retention in the lower River Spree (Germany). *Water Research* **37**: 569–578.

Schutten J, Dainty J, Davy AJ. 2004. Wave-induced hydraulic forces on submerged aquatic plants in shallow lakes. *Annals of Botany* **93**: 333–341.

Seed R. 1996. Patterns of biodiversity in the macro-invertebrate fauna associated with mussel patches on rocky shores. *Journal of the Marine Biological Association, UK* **76**: 203–210.

Seilacher A, Buatois LA, Mangano MG. 2005. Trace fossils in the Ediacaran-Cambrian transition:behavioural diversification, ecological turnover and environmental shift. *Palaeo-geography, Palaeoclimatology, Palaeoecology* **227**: 323–356.

Sekar R, Nair KVK, Rao VNR, Venugopalan VP. 2002. Nutrient dynamics and successional changes in a lentic freshwater biofilm. *Freshwater Biology* **47**: 1893–1907.

Seraphin A, Guyenne P. 2008. A flume experiment on the adjustment of the mean and turbulent statistics to a transition from short to tall sparse canopies. *Boundary-Layer Meteorology* **129**: 47–64.

Serra A, Guash H, Marti E, Geiszinger A. 2009. Measuring in-stream retention of copper by means of constant-rate additions. *Science of the Total Environment* **407**: 3847–3854.

Serra T, Fernando HJS, Rodriguez RV. 2004. Effects of emergent vegetation on lateral diffusion in wetlands. *Water Research* **38**: 139–147.

Sgro L, Mistri M, Widdows J. 2005. Impact of the infaunal Manila clam, *Ruditapes philippi-narum*, on sediment stability. *Hydrobiologia* **550**: 175–182.

Sharpe RG, James CS. 2006. Deposition of sediment from suspension in emergent vegetation. *Water SA* **32**: 211–218.

Sheldon F, Walker KF. 1997. Changes in biofilms induced by flow regulation could explain extinc-tions of aquatic snails in the lower River Murray, Australia. *Hydrobiologia* **347**: 97–108.

Shen C, Lemmin U. 1997. Ultrasonic scattering in highly turbulent clear water flow. *Ultrasonics* **35**: 57–64.

Shi Z, Hughes JMR. 2002. Laboratory flume studies of microflow environments of aquatic plants. *Hydrological Processes* **16**: 3279–3289.

Shi Z, Pethick JS, Burd F, Murphy B. 1996. Velocity profiles in a salt marsh canopy. *Geo-Marine Letters* **16**: 319–323.

Shi Z, Pethick JS, Pye K. 1995. Flow structure in and above the various heights of a saltmarsh canopy: a laboratory flume study. *Journal of Coastal Research* **11**: 1204–1209.

Shimizu Y, Tsujimoto T. 1994. Numerical analysis of turbulent open-channel flow over a vegetation layer using a k-e turbulence model. *Journal of Hydroscience and Hydraulic Engineering* **11**: 57–67.

Shiraishi F, Bissett A, de Beer D, Reimer A, Arp G. 2008a. Photosynthesis, respiration and exopolymer calcium-binding in biofilm calcification (Westerhöfer and Deinschwanger Creek, Germany). *Geomicrobiology Journal* **25**: 83–94.

Short FT, Carruthers TJB, Dennison WC, Waycott M. 2007. Global seagrass distribution and diversity: A bioregional model. *Journal of Experimental Marine Biology and Ecology* **350**: 3–20.

Siebert T, Branch GM. 2006. Ecosystem engineers: Interaction between eelgrass *Zostera capensis* and the sandprawn *Callianassa kraussi* and their indirect effects on the mudprawn *Upogebia africana*. *Journal of Experimental marine Biology and Ecology* **338**: 253–270.

Simon A, Collison AJC. 2002. Quantifying the mechanical and hydrologic effects of riparian vegetation on streambank stability. *Earth Surface Processes and Landforms* **27**(5): 527–546.

Singer G, Besemer K, Schmitt-Kopplin P, Hödl I, Battin TJ. 2010. Physical heterogeneity increases biofilm resource use and its molecular diversity in stream mesocosms. *PLoS ONE* **5**: e9988.

Smith CR, Berelson W, Demaster DJ, Dobbs FC, Hammond D, Hoover DJ, Pope RH, Stephens M. 1997. Latitudinal variations in benthic processes in the abyssal equatorial Pacific: control by biogenic particle flux. *Deep-Sea Research II.* **44**: 2295–2317

Smith CR, Jumars PA, De Master DJ. 1986. In situ studies of megafaunal mounds indicate rapid sediment turnover and community response at the deep-sea floor. *Nature* **323**: 251–253.

Smith FA, Walker NA. 1980. Photosynthesis by aquatic plants: effects of unstirred layers in relation to assimilation of CO^2 and HCO^{3-} and to carbon isotopic discrimination. *New Phytologist* **86**: 245–259.

Smith JN, Schafer CT. 1984. Bioturbation processes in continental slope and rise sediments delineated by Pb-210, microfossil and textural indicators. *Journal of Marine Research* **42**: 1117–1145.

Smith M, Vericat D, Gibbins G. 2012. Through-water terrestrial laser scanning of gravel beds at the patch scale. *Earth Surface Processes and Landforms*.

Snelgrove PVR, Butman CA, Grassle JP. 1993. Hydrodynamic enhancement of larval settlement in the bivalve *Mulinia lateralis* (Say) and the polychaete *Capitella* sp. in microdepositional environments. *Journal of Experimental Marine Biology and Ecology* **168**: 71–109.

Snover ML, Commito JA. 1998. The fractal geometry of *Mytilus edulis* L. spatial distribution in a soft-bottom system. *Journal of Experimental Marine Biology and Ecology* **223**: 53–64.

Sobral P, Widdows J. 2000. Effects of increasing current velocity, turbidity and particle-size selection on the feeding activity and scope for growth of *Ruditapes decussatus* from Ria Formosa, southern Portugal. *Journal of Experimental Marine Biology and Ecology* **245**: 111–125.

Solan M, Cardinale BJ, Downing AL, Engelhardt KAM, Ruesink JL, Srivastava DS. 2004. Extinction and ecosystem function in the marine benthos. *Science* **306**: 1177–1180.

Song T, Graf WH. 1996. Velocity and turbulence distribution in unsteady open-channels flows. *Journal of Hydraulic Engineering, ASCE* **122**: 141–154.

Song T, Graf WH, Lemmin U. 1994. Uniform flow in open channels with movable gravel bed. *Journal of Hydraulic Research* **32**: 861–876.

Souliotis D, Prinos P. 2009. Effect of a vegetation patch on turbulent channel flow. *Proceedings of the 33rd IAHR congress: water engineering for a sustainable environment* pp. 3833–3839.

Soulsby RL. 1980. Selecting record length and digitization rate for near-bed turbulence measurements. *Journal of Physical Oceanography* **10**: 208–219.

Squires GL. 1968. *Practical physics*. McGraw-Hill, London, UK.

Stamhuis EJ, Reede-Dekker T, van Etten Y, de Wiljes JJ, Videler JJ. 1996. Behaviour and time allocation of the burrowing shrimp *Callianassa subterranea* (Decapoda, Thalassinidea). *Journal of Experimental Marine Biology and Ecology* **204**: 225–239.

Stamhius EJ, Schreurs CE, Videler JJ. 1997. Burrow architecture and turbative activity of the thalassinid shrimp *Callianassa subterranea* from the central North Sea. *Marine Ecology Progress Series* **151**: 155–163.

Stamhuis EJ, Videler JJ. 1998a. Burrow ventilation in the tube dwelling shrimp *Callianassa subteranea* (*decapoda, thalassinidea*). I. *Journal of Experimental Biology* **201**: 2151–2158.

Stamhuis EJ, Videler JJ. 1998b. Burrow ventilation in the tube dwelling shrimp *Callianassa subteranea* (*decapoda, thalassinidea*). II. *Journal of Experimental Biology* **201**: 2159–2170.

Stamhuis EJ, Videler JJ. 1998c. Burrow ventilation in the tube dwelling shrimp *Callianassa subteranea* (*decapoda, thalassinidea*). III. *Journal of Experimental Biology* **201**: 2171–2181.

Stamhuis EJ, Videler JJ, van Duren LA, Müller UK. 2002. Applying digital particle image velocimetry to animal-generated flows: Traps, hurdles and cures in mapping steady and unsteady flows in Re regimes between 10^{-2} and 10^5. *Experiments in Fluids* **33**: 801–813.

Stanley SM. 1981. Infaunal survival: alternative functions of shell ornamentation in the Bivalvia (Mollusca). *Paleobiology* **7**: 384–393.

Statzner B, Arens M-F, Champagne J-Y, Morel R, Herouin E. 1999. Silk-producing stream insects and gravel erosion: significant biological effects on critical shear stress. *Water Resources Research* **35**: 3495–3506.

Statzner B, Fievet E, Champagne J-Y, Morel R, Herouin E. 2000. Crayfish as geomorphic agents and ecosystem engineers: biological behavior affects sand and gravel erosion in experimental streams. *Limnology and Oceanography* **45**: 1030–1045.

Statzner B, Fuchs U, Higler LWG. 1996. Sand erosion by mobile predaceous stream insects: implications for ecology and hydrology. *Water Resources Research* **32**: 2279–2287.

Statzner B, Lamouroux N, Nikora V, Sagnes P. 2006. The debate about drag and reconfiguration of freshwater macrophytes: comparing results obtained by three recently discussed approaches. *Freshwater Biology* **51**: 2173–2183.

Statzner B, Müller R. 1989. Standard hemispheres as indicators of flow characteristics in lotic benthos research. *Freshwater Biology* **21**: 445–459.

Statzner B, Peltret O, Tomanova S. 2003. Crayfish as geomorphic agents and ecosystem engfineers: effect of a biomass gradient on baseflow and flood-induced transport of gravel and sand in experimental streams. *Freshwater Biology* **48**: 147–163.

Statzner B, Sagnes P. 2009. Crayfish and fish as bioturbators of streambed sediments: assessing joint effects of species with different mechanistic abilities. *Geomorphology* **93**: 267–287.

Stephan U, Gutknecht D. 2002. Hydraulic resistance of submerged flexible vegetation. *Journal of Hydrology* **269**: 27–43.

Stevenson RJ. 1983. Effects of current and conditions simulating autogenically changing microhabitats on benthic diatom immigration. *Ecology* **64**: 1514–1524.

Stevenson RJ. 1996. An introduction to algal ecology in freshwater benthic habitats. In: Stevenson RJ, Bothwell ML, Lowe RL. (eds.) *Algal Ecology: Freshwater Benthic Ecosystems*. Academic, San Diego, USA. pp. 3–30.

Stewart HL. 2006. Hydrodynamic consequences of flexural stiffness and buoyancy for seaweeds: a study using physical models. *Journal of Experimental Biology* **209**: 2170–2181.

Stewart RJ, Weaver JC, Morse DE, Waite JH. 2006. The tube cement of *Phragmatopoma californica*: a solid foam. *The Journal of Experiment Biology* **207**: 4727–4734.

Stoesser T, Salvador GP, Rodi W, Diplas P. 2009. Large eddy simulation of turbulent flow through submerged vegetation. *Transport in Porous Media* **78**: 347–365.

Stone BM, Shen HT. 2002. Hydraulic resistance of flow in channels with cylindrical roughness. *Journal of Hydraulic Engineering, ASCE* **128**: 500–506.

Stone M, Krishnappan BG, Emelko MB. 2008. The effect of bed age and shear stress on the particle morphology of eroded cohesive river sediment in an annular flume. *Water Research* **42**: 4179–4187.

Stoodley P, Sauer K, Davies DG, Costerton JW. 2002. Review. Biofilms as complex differentiated communities. *Annual Review of Microbiology* **56**: 187–209.

Straatsma MW, Warmink JJ, Middelkoop H. 2008. Two novel methods for field measurements of hydrodynamic density of floodplain vegetation using terrestrial laser scanning and digital parallel photography. *International Journal of Remote Sensing* **29**: 1595–1617.

Strom KB, Papanicolaou AN. 2007. ADV measurements around a cluster microform in a shallow mountain stream. *Journal of Hydraulic Engineering, ASCE* **133**: 1379–1389.

Su XL, Li CW. 2002. Large eddy simulation of free surface turbulent flow in partly vegetated open channels. *International Journal for Numerical Methods in Fluids* **39**: 919–937.

Suchanek TH. 1979. *The Mytilus californianus community: studies on the composition, structure, organization and dynamics of a mussel bed*. PhD Thesis, University of Washington, USA.

Suchanek TH. 1983. Control of seagrass communities and sediment distribution by *Callianassa* (Crustacea, Thalassinidea) bioturbation. *Journal of Marine Research* **41**: 281–298.

Suchanek TH. 1985. Thalassinid shrimp burrows: Ecological significance of species-specific architecture. *Proceeding of the 5th International Coral Reef Congress, Tahiti* **5**: 205–210.

Suchanek TH. 1992. Extreme biodiversity in the marine environment: mussel bed communities of *Mytilus californianus*. *Northwest Environmental Journal* **8**: 150–152.

Suchanek TH, Colin PL. 1986. Rates and effects of bioturbation by invertebrates and fishes at Enewetak and Bikini atolls. *Bulletin of Marine Science* **38**: 25–34.

Suchanek TH, Colin PL, McMurtry GM, Suchanek CS. 1986. Bioturbation and redistribution of sediment radionuclides in Enewetak atoll lagoon by Callianassid shrimp: biological aspects. *Bulletin of Marine Science* **38**: 144–154.

Sukhodolov AN, Rhoads BL. 2001. Field investigation of three-dimensional flow structure at stream confluences: 2. Turbulence. *Water Resources Research* **37**: 2411–2424.

Sutherland JW. 2001 Biofilm exopolysaccharides: a strong and sticky framework. *Microbiology* **147**: 3–9.

Sutherland TF, Grant J, Amos CL. 1998. The effect of carbohydrate production by the diatom *Nitzschia curvilineata* on the erodibility of sediment. *Limnology and Oceanography* **43**: 65–72.

Sutherland, JW. 2001 Biofilm exopolysaccharides: a strong and sticky framework. *Microbiology* **147**: 3–9.

Swift DJ. 1993. The macrobenthic infauna off Sellafield (North Eastern Irish Sea) with special reference to bioturbation. *Journal of the Marine Biological Association UK* **73**: 143–162.

Swinbanks DD, Luternauer JL. 1987. Burrow distribution of thalassinidean shrimp on a Fraser Delta tidal flat, British Columbia. *Journal of Palaeontology* **61**: 315–332.

Szmeja J, Bazydlo E. 2005. The effect of water conditions on the phenology and age structure of *Luronium natans* (L.) raf. populations. *Acta Societatis Botanicorum Poloniae* **74**: 253–262.

Takao A, Kawaguchi Y, Minagawa T, Kayaba Y, Morimoto Y. 2008. The relationships between benthic macroinvertebrates and biotic and abiotic environmental characteristics downstream of the Yahagi dam, central Japan, and the state change caused by inflow from a tributary. *River Research and Applications* **24**: 580–597.

Takeda Y. 1991. Development of an Ultrasound Velocity Profile Monitor. *Nuclear Engineering and Design* **126**: 277–284.

Tal M, Paola C. 2007. Dynamic single-thread channels maintained by the interaction of flow and vegetation. *Geology* **35**: 347–350.

Tal M, Paola C. 2010. Effects of vegetation on channel morphodynamics: results and insights from laboratory experiments. *Earth Surface Processes and Landforms* **35**: 1014–1028.

Tanino Y, Nepf HM. 2008. Laboratory investigation of mean drag in a random array of rigid, emergent cylinders. *Journal of Hydraulic Engineering* **134**: 34–41.

Tao W, Hall KJ, Duff SJB. 2006. Performance evaluation and effects of hydraulic retention time and mass loading rate on treatment of woodwaste leachate in surface-flow constructed wetlands. *Ecological Engineering* **26**: 252–265.

Tao W, Hall KJ, Duff SJB. 2007a. Microbial biomass and heterotrophic production of surface flow mesocosm wetlands treating woodwaste leachate: Responses to hydraulic and organic loading and relations with mass reduction. *Ecological Engineering* **31**: 132–139.

Tao W, Hall KJ, Ramey W. 2007b. Effects of influent strength on microorganisms in surface flow mesocosm wetlands. *Water Research* **41**: 4557–4565.

Taskinen J, Saarinen M. 2006. Burrowing behaviour affects *Paraergasilus rylovi* abundance in *Anodonta piscinalis*. *Parasitology* **133**: 623–629.

Taylor GI. 1935. Statistical theory of turbulence. *Proceedings of the Royal Society of London A* **151**: 421–444.

Taylor JR. 1997. *An introduction to error analysis: the study of uncertainties in physical measurements*. (2nd edition). University Science Books, Sausalito, California, USA.

Teal LR, Bulling MT, Parker ER, Solan M. 2008. Global patterns of bioturbation intensity and mixed depth of marine soft sediments. *Aquatic Biology* **2**: 207–218.

Temmerman S, Bouma TJ, Govers G, Wang ZB, de Vries MB, Herman PMJ. 2005. Impact of vegetation on flow routing and sedimentation patterns: Three-dimensional modeling for a tidal marsh. *Journal of Geophysical Research* **110**: F04019.

ten Brinke WBM, Augustinus PGEF, Berger GW. 1995. Fine-grained sediment deposition on mussel beds in the Oosterschelde (The Netherlands), determined from echosoundings, radio-isotopes and biodeposition field experiments. *Estuarine, Coastal and Shelf Science* **40**: 195–217.

Tennekes H, Lumley JL. 1972. *A first course in turbulence*. The MIT Press, Cambridge, Massachusetts, USA.

Terrados J, Duarte CM. 2000. Experimental evidence of reduced particle resuspension within a seagrass (*Posidonia oceanica* L.) meadow. *Journal of Experimental Marine Biology and Ecology* **243**: 45–53.

Thayer CW. 1979. Biological bulldozers and the evolution of marine benthic communities. *Science* **203**: 458–461.

Thieltges DW, Buschbaum C. 2006. Mechanism of an epibont burden: *Crepidula fornicata* increases byssus thread production by *Mytilus edulis*. *Journal of Molluscan Studies* **73**: 75–77.

Thomas RE, Pollen-Bankhead N. 2010. Modeling root-reinforcement with a fiber-bundle model and Monte Carlo simulation. *Ecological Engineering* **36**(1): 47–61.

Thomason JC, Hills JM, Clare AS, Neville A, Richardson M. 1998. Hydrodynamic consequences of barnacles colonization. *Hydrobiologia* **375/376**: 191–201.

Thompson CEL, Amos CL. 2002. The impact of mobile disarticulated shells of *Cerastoderma edulis* on the abrasion of a cohesive substrate. *Estuaries* **25**: 204–214.

Thompson CEL, Amos CL, Umgiesser G. 2004. A comparison between fluid shear stress reduction by halophytic plants in Venice Lagoon, Italy and Rustico Bay, Canada – analyses of *in situ* measurements. *Journal of Marine Systems* **51**: 293–308.

Thomson JR, Clark BD, Fingerut JT, Hart DD. 2004. Local modification of benthic flow environments by suspension-feeding stream insects. *Oecologia* **140**: 533–542.

Thrush SF, Hewitt JE, Gibbs MM, Lundquist C, Norkko A. 2006. Functional role of large organisms in intertidal communities: community effects and ecosystem function. *Ecosystems* **9**: 1029–1040.

Timoshenko S. 1955. *Strength of materials; Part 1: Elementary theory and problems*. Van Nostrand, New York.

Tinoco RO, Cowen EA. 2009. Experimental study of flow through macrophyte canopies. In: *Water Engineering for a Sustainable Environment, Proceedings of the 33rd IAHR Congress, Vancouver BC, Canada, August 9–14, 2009*. pp. 6160–6166. IAHR: Madrid, Spain.

Tinoco RO, Cowen EA. 2013. The direct and indirect measurement of boundary stress and drag on individual and complex arrays of elements. *Experiments in Fluids* 54(4): 1509.

Tiselius P, Hansen B, Jonsson P, Kiorboe T, Nielsen G, Piontkovski S, Saiz E. 1995. Can we use laboratory-reared copepods for experiments? A comparison of feeding behaviour and reproduction between a field and a laboratory population of *Acartia tonsa*. *Journal of Marine Science* 52: 369–376.

Tlili A, Montuelle B, Bérard A, Bouchez A. 2011. Impact of chronic and acute pesticide exposures on periphyton communities. *Science of the Total Environment* 409: 2102–2113.

Tooth S, Nanson GC. 1999. Anabranching rivers on the Northern Plains of arid central Australia. *Geomorphology* 29: 211–233.

Tooth S, Nanson GC. 2000. The role of vegetation in the formation of anabranching channels in an ephemeral river, Northern plains, arid central Australia. *Hydrological Processes* 14(16–17), 3099–3117.

Townsin RL. 2003. The ship hull fouling penalty. *Biofouling* 19: 9–15.

Trager GC, Hwang J-S, Strickler JR. 1990. Barnacle suspension-feeding in variable flow. *Marine Biology* 105: 117–127.

Troost K, Stamhuis EJ, van Duren LA, Wolff WJ. 2009. Feeding current characteristics of three morphologically different bivalve suspension feeders, *Crassotrea gigas*, *Mytilus edulis* and *Cerastoderma edule*, in relation to food competition. *Marine Biology* 156: 355–372.

Tsujimoto T. 1999. Fluvial processes in streams with vegetation. *Journal of Hydraulic Research, IAHR*: 37: 789–803

Tsujimoto T, Shimizu Y, Kitamura T, Okada T. 1992. Turbulent open-channel flow over bed covered by rigid vegetation. *Journal of Hydroscience and Hydraulic Engineering, JSCE* 10: 13–26.

Türker U, Yagci O, Kabdasll MS. 2006. Analysis of coastal damage of a beach profile under the protection of emergent vegetation. *Ocean Engineering* 33: 810–828.

Uehlinger U, Buhrer H, Reichert P. 1996. Periphyton dynamics in a flood prone pre alpine river: Evaluation of significant processes by modelling. *Freshwater Biology* 36: 249–263.

Uncles RJ. 2002. Estuarine physical processes research: Some recent studies and progress. *Estuarine, Coastal and Shelf Science* 55: 829–856.

Underwood GJC, Paterson DM. 1993. Seasonal changes in diatom biomass, sediment stability and biogenic stabilization in the Severn estuary. *Journal of the Marine Biological Association of the UK* 73: 871–887.

Underwood GJC, Perkins RG, Consalvey MC, Hanlon ARM, Oxborough K, Baker NR, Paterson DM. 2005. Patterns in microphytobenthic primary productivity: Species-specific variation in migratory rhythms and photosynthetic efficiency in mixed-species biofilms. *Limnology and Oceanography* 50: 755–767.

van de Koppel J, Gascoigne JC, Theraulaz G, Rietkerk M, Mooij WM, Herman PMJ. 2008. Experimental evidence for spatial self-organization and its emergent effects in mussel bed ecosystems. *Science* 322: 739–742.

van de Koppel J, Rietkerk M, Dankers N, Herman PMJ. 2005. Scale-dependent feedback and regular spatial patterns in young mussel beds. *The American Naturalist* 165: E66–E77.

van de Lageweg WI, van Dijk WM, Hoendervoolgt R, Kleinhans MG. 2010. *Effects of riparian vegetation on experimental channel dynamics*. Proceedings of the International Conference on Fluvial Hydraulics River Flow 2010. Braunschweig, Germany.

van Duren LA, Herman PMJ, Sandee AJJ, Heip CHR. 2006. Effects of mussel filtering activity on boundary layer structure. *Journal of Sea Research* 55: 3–14.

van Duren LA, Stamhuis EJ, Videler JJ. 2003. Copepod feeding currents: flow patterns, filtration rates and energetics. *Journal of Experimental Biology* 206: 255–267 van Leeuwen B, Augustijn DCM, van Wesenbeeck BK, Hulscher SJMH, de Vries MB. 2010. Modelling the influence of a young mussel bed on fine sediment dynamics on an intertidal flat in the Wadden Sea. *Ecological Engineering* **36**: 145–153.

van Proosdij D, Davidson-Arnott RGD, Ollerhead J. 2006. Controls on spatial patterns of sediment deposition across a macro-tidal salt marsh surface over single tidal cycles. *Estuarine, Coastal and Shelf Science* **69**: 64–86.

van Wesenbeeck *et al.* 2008. Does scale-dependent feedback explain spatial complexity in salt-marsh ecosystems. *Oikos* **117**: 152–159.

Vaughn CC, Hakenkamp CC. 2001. The functional role of burrowing bivalves in freshwater ecosystems. *Freshwater Biology* **46**: 1431–1446.

Vedel A, Andersen BB, Riisgård HU. 1994. Field investigations of pumping activity of the facultatively filter-feeding polychaete *Nereis diversicolor* using an improved infrared phototransducer system. *Marine Ecology Progress Series* **103**: 91–101.

Velasco D, Bateman A, Medina V. 2008. A new integrated, hydro-mechanical model applied to flexible vegetation in riverbeds. *Journal of Hydraulic Research* **46**: 579–597.

Ventura M, Liboriussen L, Lauridsen T, Søndergaard M, Søndergaard M, Jeppesen E. 2008. Effects of increased temperature and nutrient enrichment on the stoichiometry of primary producers and consumers in temperate shallow lakes. *Freshwater Biology* **53**: 1434–1452.

Verduin JJ, Backhaus JO. 2000. *Dynamics of plant-flow interactions for the seagrass* Amphibolis antarctica: *Field observations and model simulations. Estuarine, Coastal and Shelf Science* **50**: 185–204.

Verduin JJ, Backhaus JO, Walker DI. 2002. Estimates of pollen dispersal and capture within *Amphibolis antarctica* (Labill.) sonder and aschers. ex aschers. meadows. *Bulletin of Marine Science* **71**: 1269–1277.

Vermaat JE, Santamaria L, Roos PJ. 2000. Water flow across and sediment trapping in submerged macrophyte beds of contrasting growth form. *Archiv Fur Hydrobiologie* **148**: 549–562.

Verwey J. 1952. On the ecology of distribution of cockle and mussel in the Dutch Waddenzee, their role in sedimentation, and source of their food supply, with a short review of the feeding behaviour of bivalve molluscs. *Archives Néerlandaises de Zoologie* **10**: 172–239.

Villeneuve A, Montuelle B, Bouchez A. 2011. Effects of flow regime and pesticides on periphytic communities: Evolution and role of biodiversity. *Aquatic Toxicology* **102**: 123–133.

Vincent B, Desrosiers G, Gratton Y. 1988. Orientation of the infaunal bivalve *Mya arenaria* L. in relation to local current direction on a tidal flat. *Journal of Experimental Marine Biology and Ecology* **124**: 205–214.

Vionnet CA, Tassi PA, Martin Vide JP. 2004. Estimates of flow resistance and eddy viscosity coefficients for 2D modelling on vegetated floodplains. *Hydrological processes* **18**: 2907–2926.

Vogel S. 1994. *Life in Moving Fluids*. 2nd Edition. Princeton University Press. USA.

Vogel S, Bretz WL. 1971. Interfacial organisms: passive ventilation in the velocity gradients near surfaces. *Science* **175**: 210–211.

Volkenborn N, Polerecky L, Wethey DS, Woodin SA. 2010. Oscillatory porewater bioadvection in marine sediments induced by hydraulic activities of *Arenicola marina*. *Limnology and Oceanography* **55**: 1231–1247.

Volkenborn N, Reise K. 2006. Lugworm exclusion experiment: responses by deposit feeding worms to biogenic habitat transformations. *Journal of Experimental Marine Biology and Ecology* **330**: 169–179.

Voulgaris G, Meyers ST. 2004. Net effect of rainfall activity on salt-marsh sediment distribution. *Marine Geology* **207**: 115–129.

Voulgaris G, Trowbridge JH. 1998. Evaluation of the acoustic Doppler velocimeter (ADV) for turbulence measurements. *Journal of Atmospheric and Oceanic Technology* **15**: 272–289.

Vu B, Chen M, Russell JC, Ivanova EP. 2009. Bacterial Extracellular Polysaccharides Involved in Biofilm Formation. *Molecules* **14**: 2535–2554.

Wagenschein D, Rode M. 2008. Modelling the impact of river morphology on nitrogen retention—A case study of the Weisse Elster River (Germany). *Ecological Modelling* **211**: 224–232.

Wahl M. 1996. Fouled snails in flow: potential of epibionts on *Littorina littorea* to increase drag and reduce snail growth rates. *Marine Ecology Progress Series* **138**: 157–168.

Wahl M, Hay ME, Enderlein P. 1997. Effects of epibiosis on consumer-prey interactions *Hydrobiologia* **355**: 49–59.

Wahl M, Shahnaz L, Dobretsov S, Saha M, Symanowski F, David K, Lachnit T, Vasel M, Weinberger F. 2010. Ecology of antifouling resistance in the bladder wrack Fucus vesiculosus: patterns of microfouling and antimicrobial protection. *Marine Ecology Progress Series* **411**: 33–48.

Wahl TL. 2003. Discussion of "Despiking Acoustic Doppler Velocimeter Data" by Derek G. Goring and Vladimir I. Nikora. *Journal of Hydraulic Engineering, ASCE* **129**: 484–487.

Waite JH. 2002. Adhesion á la Moule. *Integrative and Comparative Biology* **42**: 1172–1180.

Waite JH. Vaccaro E, Sun C, Lucas JM. 2002. Elastomeric gradients: a hedge against stress concentration in marine holdfast? *Philosophical Transactions of the Royal Society B* **357**: 143–153.

Waldron LJ. 1977. The shear resistance of root-permeated homogeneous and stratified soil. *Soil Science Society of America Journal* **41**(5), 843–849.

Waldron LJ, Dakessian S. 1981. Soil reinforcement by roots: calculation of increased soil shear resistance from root properties. *Soil Science* **132**(6), 427–435.

Wallace S. Luketina D, Cox R. 1998. Large scale turbulence in seagrass canopies. In: *Proceedings of the 13th Australasian Fluid Mechanics Conference, Monash Unviersity, Melbourne, Australia, 13–18 December 1998*. pp. 973–976.

Wang, J. Chen, C. 2009. Biosorpents for heavy metal removal and their future. *Biotechnology Advances* **27**: 195–226

Warburton K. 1976. Shell form, behaviour, and tolerance to water movement in the limpet *Patina pellucida* (L.) (Gastropoda: Prosobranchia). *Journal of Experimental Marine Biology and Ecology* **23**: 307–325.

Ward LG, Kemp WM, Boynton WR. 1984. The influence of waves and seagrass communities on suspended particles in an estuarine embayment. *Marine Geology* **59**: 85–103.

Warwick RM, McEvoy AJ, Thrush SF. 1997. The influence of *Atrina zelandica* Gray on meiobenthic nematode diversity and community structure. *Journal of Experimental Marine Biology and Ecology* **214**: 231–247.

Watermanm F, Hillebrand H, Gerdes G, Krumbein WE, Sommer U. 1999. Competition between benthic cyanobacteria and diatoms as influenced by different grain sizes and temperatures. *Marine Ecology Progress Series* **187**: 77–87.

Watters GT, O'Dee SH, Chordas III, S. 2001. Patterns of vertical migration in freshwater mussels (Bivalvia: Unionoida). *Journal of Freshwater Ecology* **16**: 541–549.

Wayne CJ. 1976. The effects of sea and marsh grass on wave energy. *Coastal Research Notes* **4**: 6–8.

Welker M, Walz N. 1998. Can mussels control the plankton in rivers? – a planktological approach applying a Lagrangian sampling strategy. *Limnology and Oceanography* **43**: 753–762.

Westerweel J. 1994. Efficient detection of spurious vectors in particle image velocimetry data. *Experiments in Fluids* **16**: 236–247.

Westerweel J. 1997. Fundamentals of Digital Particle Image Velocimetry. *Measurement Science and Technology* **8**: 1379–1392.

Wheatcroft RA. 1992. Experimental tests for particle size-dependent bioturbation in the deep ocean. *Limnology and Oceanography* 37: 90–104.

Wheatcroft RA, Olmex I, Pink FX. 1994. Particle bioturbation in Massachusetts Bay: preliminary results using a new deliberate tracer technique. *Journal of Marine Research* 52: 1129–1150.

White BL, Nepf HM. 2003. Scalar transport in random cylinder arrays at moderate Reynolds number. *Journal of Fluid Mechanics* 487: 43–79.

White BL, Nepf HM. 2007. Shear instability and coherent structures in shallow flow adjacent to a porous layer. *Journal of Fluid Mechanics* 593: 1–32.

White BL, Nepf HM. 2008. A vortex-based model of velocity and shear stress in a partially vegetated shallow channel. *Water Resources Research* 44: W01412, DOI:10.1029/2006 WR005651.

Widdows J, Blauw A, Heip CHR, Herman PMJ, Lucas CH, Middelburg JJ, Schmidt S, Brinsley MD, Twisk F, Verbeek H. 2004. Role of physical and biological processes in sediment dynamics of a tidal flat in Westerschelde Estuary, SW Netherlands *Marine Ecology Progress Series* 274: 41–56

Widdows J, Brinsley MD, Salkeld PN, Elliot M. 1998a. Use of annular flumes to determine the influence of current velocity and bivalves on material flux at the sediment-water interface. *Estuaries* 21: 552–559.

Widdows J, Brinsley MD, Bowley N, Barrett C. 1998b. A benthic annular flume for *in situ* measurement of suspension feeding/biodeposition rates and erosion potential of intertidal cohesive sediments. *Estuarine, Coastal and Shelf Science* 46: 27–38.

Widdows J, Brinsley, MD, Salkeld PN, Lucas CH. 2000a. Influence of biota on spatial and temporal variation in sediment erodability and material flux on a tidal flat (Westerschelde, The Netherlands). *Marine Ecology Progress Series* 194: 23–37.

Widdows J, Brown S, Brinsley MD, Salkeld PN, Elliot M. 2000b. Temporal changes in intertidal sediment erodability: influence of biological and climatic factors. *Continental Shelf Research* 20: 1275–1289.

Widdows J, Lucas JS, Barrett C, Brinsely MD, Salkeld PN, Staff FJ. 2002. Investigation of the effects of current velocity on mussel feeding and mussel bed stability using an annular flume. *Helgolander Wissenschaftliche Meeresuntersuchungen* 56: 3–12.

Widdows J, Navarro JM. 2007. Influence of current speed on clearance rate, algal cell depletion in the water column and resuspension of biodeposits of cockles (*Cerastoderma edule*). *Journal of Experimental Marine Biology and Ecology* 343: 44–51.

Widdows J, Pope ND, Brinsley MD, Asmus H, Asmus R. 2008a. Effect of seagrass beds (*Zostera noltii* and *Z. marina*) on near-bed hydrodynamics and sediment resuspension. *Marine Ecology Progress Series* 358: 125–136.

Widdows J, Pope N, Brinsley M. 2008b. Effect of *Spartina anglica* stems on nearbed hydrodynamics, sediment erodability and morphological changes on an intertidal mudflat. *Marine Ecology Progress Series* 362: 45–57.

Widdows J, Pope ND, Brinsley MD, Gascoigne J, Kaiser MJ. 2009. Influence of self-organised structures on near-bed hydrodynamics and sediment dynamics within a mussel (*Mytilus edulis*) bed in the Menai Strait. *Journal of Experimental Marine Biology and Ecology* 379: 92–100.

Wienke J, Oumeraci H. 2005. Breaking wave impact force on a vertical and inclined slender pile: theoretical and large-scale model investigations. *Coastal Engineering* 52: 435–462.

Wiggins GB. 2004. *Caddisflies: the underwater architects*. University of Toronto Press, USA.

Wilcock PR. 1996. Estimating local bed shear stress from velocity observations. *Water Resources Research* 32: 3361–3366.

Wildish D, Kristmanson D. 1997. *Benthic Suspension Feeders and Flow*. Cambridge University Press, Cambridge, UK.

Wildish DJ, Kristmanson DD, Hoar RL, DeCoste AM, McCormick SD, White AW. 1987. Giant scallop feeding and growth responses to flow. *Journal of Experimental Marine Biology and Ecology* 113: 207–220.

Wildish DJ, Miyares MP. 1990. Filtration rate of blue mussels as a function of flow velocity: preliminary experiments. *Journal of Experimental Marine Biology and Ecology* 142: 213–219.

Willows RI, Widdows J, Wood RG. 1998. Influence of an infaunal bivalve on the erosion of an intertidal cohesive sediment: a flume and modelling study. *Limnology and Oceanography* 43: 1332–1343.

Wilson CAME, Horritt MS. 2002. Measuring the flow resistance of submerged grass. Hydrological Processes 16: 2589–2598.

Wilson CAME, Stoesser T, Bates PD, Batemann Pinzen A. 2003. Open channel flow through different forms of submerged flexible vegetation. Journal of Hydraulic Engineering, ASCE 129: 847–853.

Wilson CAME, Stoesser T, Bates PD. 2005. Modelling of open channel flow through vegetation. In: Bates PD, Lane SN, Ferguson RI. (eds) Computational Fluid Dynamics: Applications in Environmental Hydraulics. John Wiley & Sons, Chichester, UK. pp. 395–428.

Wilson CAME, Xavier P, Schoneboom T, Aberle J, Rauch H-P, Lammeranner W, Weissteiner C, Thomas H. 2010. The hydrodynamic drag of full scale trees. In: Dittrich A, Koll K, Aberle J, Geisenhainer P. (eds) River Flow 2010. Bundesanstalt für Wasserbau, Karlsruhe, Germany. pp. 453–459.

Wilson CAME, Yagci O, Rauch HP, Olsen NRB. 2006a. 3D numerical modelling of a willow vegetated river/floodplain system. Journal of Hydrology 327: 13–21.

Wilson CAME, Yagci O, Rauch H-P, Stoesser T. 2006b. Application of the drag force approach to model the flow-interaction of natural vegetation. International Journal of River Basin Management 4: 137–146.

Wilson NR, Shaw RH. 1977. A higher order closure model for canopy flow. *Journal of Applied Meteorology* 16: 1198–1205.

Wingender J, Neu TR, Flemming HC. 1999. What are bacterial extracellular polymeric substances In: Wingender J, Neu TR, Flemming HC. Eds. *Microbial extracellular polymeric substances – Characterization, Structure and Function*. New York, Springer.

Winter KG. 1977. An outline of the techniques available for the measurement of skin friction in turbulent boundary layers. *Progress in Aerospace Sciences* 18: 1–57.

Winterwerp JC, van Kesteren WGM. 2004. *Introduction to the physics of cohesive sediment dynamics in the marine environment*. Developments in Sedimentology, Vol. 56. Elsevier, Amsterdam, The Netherlands.

Witman JD, Suchanek TH. 1984. Mussels in flow: drag and dislodgement by epizoans. *Marine Ecology Progress Series* 16: 259–268.

Wolanski E. 1995. Transport of sediment in mangrove swamps. *Hydrobiologia* 295: 31–42.

Wolfaardt GM, Lawrence JR, Robarts RD, Caldwell DE. 1998. In situ characterization of biofilm exopolymers involved in the accumulation of chlorinated organics. *Microbiolial Ecology* 35: 213–223.

Wood PJ, Armitage PD. 1997. Biological effects of fine sediment in the lotic environment. *Environmental Management* 21: 203–217.

Woodin SA, Wethey DS, Volkenborn N. 2010. Infaunal hydraulic ecosystem engineers: cast of characters and impacts. *Integrative and Comparative Biology* 50: 176–187.

Woodruff SL, House WA, Callow ME, Leadbeater BSC. 1999. The effects of biofilms on chemical processes in surficial sediements. *Freshwater Biology* 41: 73–89.

Wootton JT. 2001. Local interactions predict large-scale pattern in empirically derived cellular automata. *Nature* 413: 841–844.

Wotton RS. 2011. EPS (Extracellular Polymeric Substances), silk, and chitin: vitally important exudates in aquatic ecosystems. *Journal of the North American Benthological Society* 30: 762–769.

Wright LD, Schaffner LC, Maa JP-Y. 1997. Biological mediation of bottom boundary layer processes and sediment suspension in the lower Chesapeake Bay. *Marine Geology* 141: 27–50.

Wu CL, Chau KW, Li YS. 2009. Predicting monthly streamflow using data-driven models coupled with data-preprocessing techniques. *Water Resources Research* 45: W08432.

Xavier P, Wilson C, Aberle J, Rauch HP, Schoneboom T, Lammeranner W, Thomas H. 2010. Drag force of flexible submerged trees. In: Grüne J, Klein Breteler M. (eds) *Proceedings of the HYDRALAB III Joint User Meeting, Hannover, February 2010*. Deltares, Delft, The Netherlands. pp. 263–266.

Yagci O, Tschiesche U, Kabdasli MS. 2010. The role of different forms of natural riparian vegetation on turbulence and kinetic energy characteristics. *Advances in Water Resources* 33: 601–614.

Yager PL, Nowell ARM, Jumars PA. 1993. Enhanced deposition to pits: a local food source for benthos. *Journal of Marine Research* 51: 209–236.

Yallop ML, Paterson DM, Wellsbury P. 2000. Interrelationships between rates of microbial production, exopolymer production, microbial biomass, and sediment stability in biofilms of intertidal sediments. *Microbial Ecology* 39: 116–127.

Yamaguchi K. 1998. Cementation vs mobility: development of a cemented byssus and flexible mobility in *Anomia chinensis*. Marine Biology 132: 651–661.

Yonge CM. 1960. *Oysters*. Collins, London, UK.

Ysebaert T, Hart M, Herman PMJ. 2009. Impacts of bottom and suspended cultures of mussels *Mytilus* spp. on the surrounding sedimentary environment and macrobenthic biodiversity. *Helgoland Marine Research* 63: 59–74.

Yund PO, Meidel SK. 2003. Sea urchin spawning in benthic boundary layers: Are eggs fertilized before advecting away from females? *Limnology and Oceanography* 48: 795–801.

Zanetell BA, Peckarsky BL. 1996. Stoneflies as ecological engineers - hungry predators reduce fine sediments in stream beds. *Freshwater Biology* 36: 569–577.

Zehm A, Nobis M, Schwabe A. 2003. Multiparameter analysis of vertical vegetation structure based on digital image processing. *Flora* 198: 142–160.

Zhou ZC, Shangguan ZP. 2005. Soil anti-scouribility enhanced by plant roots. *Journal of Integrated Plant Biology* 47: 676–682.

Ziagova M, Dimitriadis G, Aslanidou D, Papaioannou X, Tzannetaki EL, Liakopoulou- Kyriakides M. 2007. Comparative study of Cd(II) and Cr(VI) biosorption on *Staphylococcus xylosus* and *Pseudomonas sp* in single and binary mixtures. *Bioresources Technology* 98: 2859–2865.

Ziebis W, Forster S, Huettel M, Jorgensen BB. 1996. Complex burrows of the mud shrimp *Callianassa truncata* and their geochemical impact in the sea bed. *Nature* 382: 619–622.

Zong L, Nepf H. 2010. Flow and deposition in and around a finite patch of vegetation. *Geomorphology* 116: 363–372.

Zorn ME, Lalonde SV, Gingras MK, Pemberton SG, Konshauser KO. 2006. Microscale oxygen distribution in various invertebrate burrow walls. *Geobiology* 4: 137–145.

Zrnic DS. 1977. Spectral moment estimates from correlated pulse pairs. *IEEE Transactions on Aerospace and Electronic Systems* AES–9: 151–165.